ERGEBNISSE DER MATHEMATIK UND IHRER GRENZGEBIETE

UNTER MITWIRKUNG DER SCHRIFTLEITUNG DES „ZENTRALBLATT FÜR MATHEMATIK“

HERAUSGEGEBEN VON

L. V. AHLFORS · R. BAER · R. COURANT · J. L. DOOB
S. EILENBERG · P. R. HALMOS · T. NAKAYAMA
H. RADEMACHER · F. K. SCHMIDT · B. SEGRE · E. SPERNER

NEUE FOLGE · HEFT 6

REIHE:

ALGEBRAISCHE GEOMETRIE

BESORGT VON

B. SEGRE

SPRINGER-VERLAG BERLIN HEIDELBERG GMBH
1955

ALGEBRAIC THREEFOLDS

WITH SPECIAL REGARD
TO PROBLEMS OF RATIONALITY

BY

L. ROTH

SPRINGER-VERLAG BERLIN HEIDELBERG GMBH
1955

URSPRÜNGLICH ERSCHIENEN BEI SPRINGER-VERLAG OHG. BERLIN · GÖTTINGEN · HEIDELBERG 1955

ISBN 978-3-540-01889-6 ISBN 978-3-642-85531-3 (eBook)
DOI 10.1007/978-3-642-85531-3

BRÜHLSCHE UNIVERSITÄTSDRUCKEREI GIESSEN

Preface.

The present monograph, which is based on, but does not entirely supersede, the writer's report (a), treats of algebraic threefolds from the classical algebro-geometric point of view. Since, however, classical methods at present require the help of transcendental and topological theory at several points, we have inserted a brief account of this theory wherever necessary. We have also indicated the various extensions of our results to varieties of higher dimension or, when desirable, have inverted this procedure, passing from the general to the particular.

The subtitle of this work may need some justification: it is intended to describe roughly that type of question which, in the theory of surfaces, usually but not invariably leads to the rational or scrollar surfaces as its solution, and which therefore suggests itself quite naturally in the study of threefolds. But in this field, solutions to the analogous problems are at the moment — and perhaps always will be — far less precise, and by no means lead to birational or ruled threefolds to anything like the same extent.

The detailed scheme of the work may be seen from the list of Contents which follows. As regards the general plan: in chapters I—III, which are not governed by the subtitle, those results which are established exactly as in the corresponding theory of surfaces — for which we refer the reader to Enriques (a) and Zariski (a) — have usually been stated without proof; of the rest, at least an indication of proof has been given wherever possible. On the other hand, the subject matter of chapters IV—VI has been developed as fully as space has allowed.

The above references to treatises on the theory of surfaces are intended for those who are already familiar, to some extent at least, with the subject. But since the books in this series aim at being — so far as is possible — complete in themselves, something more has been offered to the non-specialist, namely an Appendix in the shape of a brief survey of the chief concepts and theorems which underly the present work. This, we hope, will go some way towards furnishing the necessary background; at any rate, at the very least it can serve as a useful compendium of definitions. Since, however, it ranks as merely optional reading, it has been placed at the end of the book instead of at the beginning.

It remains only to add that a knowledge of the classical geometry of hyperspace is of course assumed; and that, unless the contrary is stated, all the varieties considered in the sequel are algebraic and irreducible and defined in a complex projective space; and, finally, that the Bibliography is not intended to be exhaustive, being chiefly devoted to references which have practical rather than historical interest.

London, September 1955.

LEONARD ROTH.

Contents.

Chapter I.

Invariantive Theory.

1. Ordinary singularities. Projective characters. For general notions concerning algebraic varieties reference may be made to SAMUEL (a) and HODGE and PEDOE (a); in this work we shall be mainly concerned with threefolds, which in all cases are *irreducible* and defined in *a complex projective space* S_r $(r \geqq 4)$. We shall denote such a threefold by U, V or W: surfaces lying on it will be denoted by capitals $A, B, C, \ldots$, curves by Gothic letters $\mathfrak{A}, \mathfrak{B}, \mathfrak{C}, \ldots$, and sets of points by Greek letters $\alpha, \beta, \gamma, \ldots$.

Much of the invariantive theory of threefolds requires that the threefolds under consideration should be either non-singular or endowed with *ordinary singularities* (ZARISKI [1] has shown that any threefold may be transformed birationally into one free from singularities); these are of the kind that would arise if a non-singular threefold in higher space were projected generically on to S_r $(r \leqq 6)$. Thus, if the threefold V lies in S_6, it has a finite number of improper nodes, if in S_5, an improper nodal curve (possibly reducible or of order zero); while, if V is a primal of S_4, we have the following scheme:

On V there is an ordinary double surface D (and not a surface of higher multiplicity; cf. ENRIQUES, a) which is free from improper nodes. On D there lies a curve $\mathfrak{C}$ which is an ordinary triple curve for V and for D; and on $\mathfrak{C}$ there is a finite number of points which are quadruple for V and for $\mathfrak{C}$ and sextuple for D. On D there is a cuspidal curve $\mathfrak{C}'$ (or curve of pinchpoints) which meets $\mathfrak{C}$ in a number of points which we call stationary points of V. The curve $\mathfrak{C}'$ is non-singular, and $\mathfrak{C}$ has no other multiple points than those described above.

We note that, if V is in fact a general projection, then D must be irreducible; for FRANCHETTA [1, 2] has shown that, if a surface of S_3, which is the general projection of a surface with ordinary singularities, situated in S_4, has a reducible double curve, then it must be a STEINER surface; and it will appear later that no primal having only ordinary singularities can possess a general surface section of this type.

We now introduce the following projective characters (ROTH [1, 2]): the order n of V; the elementary projective characters of D (order α_0, rank, α_1, class α_2, *ceto* β_2); the order t and rank r of $\mathfrak{C}$; the number ξ of quadruple points and the number τ of stationary points. It is also convenient to introduce temporarily the class of immersion t_1 of $\mathfrak{C}$ in V, given by the formula $t_1 = t(n-3) - 4\xi$. We then have the following equations between the characters:

$$\tau = 2(t_1 - r), \quad \alpha_0(\alpha_0 - 1) = \alpha_1 + 2t_1 + 6\xi + \beta_2,$$

the latter expressing the fact that D is without improper nodes. Thus in all we have ten projective characters connected by three relations.

Supposing instead that V is situated in S_r ($r \geqq 6$) we may define for it a set of seven *elementary projective characters* (SEVERI [2]). Of these, four are the elementary projective characters of a general prime section (order μ_0, rank μ_1, class μ_2, *ceto* ν_2), and the others are the so-called *last characters*, expressed in SCHUBERT's notation by the symbols

$$\mu_3 = (r-4, r-3, r-2, r), \quad \nu_3 = (r-6, r-2, r-1, r),$$
$$\nu_{21} = (r-5, r-3, r-1, r).$$

Of these, μ_3 and ν_3 are respectively the class and *ceto* of V. It will be seen that the restriction on r is essential, since ν_3 is not defined when $r < 6$, while ν_{21} is not defined when $r < 5$.

In the case where V is the complete intersection of $r-3$ non-singular primals, in general position, the values of μ_3 and ν_3 are known from the SEVERI formulae for intersections of varieties; and in this case ν_{21} is easily calculated (ROTH [3], TODD [1]). But no results have been obtained for the case where V is a partial intersection.

The seven characters defined above are invariant for general projection of V on to a space of dimension not less than 6; but it is desirable (and, in the sequel, necessary) to find two projective characters which, like the former, are additive, and which are defined for all $r \geqq 4$. These are suggested by the fact that, when V is projected into a primal W of S_4, the character ν_2 becomes the order of the cuspidal curve $\mathfrak{C}'$; we therefore choose for the new characters the rank ϱ of $\mathfrak{C}'$ and the class ζ of immersion of $\mathfrak{C}'$ in W; these are given by the formulae (ROTH [4])

$$\varrho = 3\nu_3 + 2\nu_{21} - 2\nu_2, \quad \zeta = 4\nu_3 + 2\nu_{21} - 2\nu_2.$$

2. Adjoint primals in S_4. It will be convenient to anticipate later developments in order to give other results of a projective nature which we shall afterwards require. Here and in § 6 the surface D is *irreducible*.

Let V be a primal, endowed with ordinary singularities and with projective characters as defined in § 1; and let U be an *adjoint primal* of given order l: that is, U is constrained to pass simply through the double surface D, doubly through the triple curve $\mathfrak{C}$, and triply through the ξ quadruple points of V. It can be shown that, if l is sufficiently large, the linear system $|U|$ defined by U cuts on V, residually to D, a system $|F^{(l)}|$ of surfaces $F^{(l)}$ which is effective, composed of irreducible surfaces, and endowed with an irreducible characteristic curve $\mathfrak{C}^{(l)}$. For such values of l we may calculate (ROTH [5, 6]) the elementary projective characters of $F^{(l)}$ and $\mathfrak{C}^{(l)}$, and hence the grade of the system $|F^{(l)}|$; we may then deduce formulae for the numerical invariants of $F^{(l)}$.

It may happen that, for an assigned value of l, one or more of the above entities is reducible or virtual; in particular, for $l = n - 5$, the primals U, which are then called the *canonical adjoints*, may be effectively non-existent. In any case the system $|F^{(n-5)}|$ is called the *virtual impure canonical system* of V, and is denoted by $|K|$; using the above results we can at once write down its virtual characters. Thus the virtual grade Ω_0 is given by

$$\Omega_0 = n(n-5)^3 - \alpha_0(9n^2 - 89n + 218) + 2(\alpha_0^2 - \alpha_2 - \beta_2) + \alpha_1(7n - 38) + \\ + 2t(7n - 66) + 16\xi + r\,.$$

Again, calculating the virtual curvilinear genus Ω_1 of $|K|$ we find that

$$2\,\Omega_1 - 2 = 3\,\Omega_0\,.$$

Further, denoting by ω the virtual linear genus, or CASTELNUOVO-ENRIQUES invariant of K, we obtain the relation

$$\omega = 4\,\Omega_0 + 1\,.$$

These last two equations are due to PANNELLI [1]; later we shall re-obtain them by another method.

3. Linear systems of surfaces. In all that follows it is to be assumed, unless the contrary is stated, that V is either a primal with ordinary singularities or a non-singular threefold in higher space. For all such varieties the introduction of linear systems of surfaces by means of rational functions, and the construction of complete linear systems by the use of adjoints, are effected precisely as in the theory of surfaces (ENRIQUES, a; ZARISKI, a). Thus, if V is a primal, the complete linear system $|A|$ characterised by a given surface A is obtainable by drawing through A an adjoint primal U of conveniently high order, and then taking the system $|U|$ of adjoints which pass through the surface B residual to A and the double surface D. If, instead, V lies in S_r ($r > 4$), we proceed as follows (SEVERI [3]): through V we draw $r-3$ primals of sufficiently high orders $n_1, n_2, \ldots$, meeting residually in a threefold V' which is irreducible and non-singular; then the primals which contain V' but not V play a role analogous to that of the adjoints in the pre vious case.

Linear surface systems on V of freedom $r > 1$ possess the characteristic property that their *index* (number of members which pass through r general points of V) is unity; for $r = 1$, however, an exception is provided by the irrational pencil, which we denote by a symbol such as $\{A\}$. Also, as in the theory of curve systems on a surface, we have BERTINI's theorems, which lead to the following classification of linear systems:

I. If the general surface of the given system is reducible, then the variable part of the system is either irreducible or is compounded of a pencil, rational or irrational.

II. If the general surface of the system is irreducible, there are three possibilities.

a) The characteristic series of the system is simple and effectively of positive order n; then, if $r \geqq 4$, it may be mapped on the prime sections of a threefold W, of order n, which is birationally equivalent to V.

b) The characteristic series is compounded of an involution, of order ν, say; again, if $r \geqq 4$, the system may be mapped on the prime sections of a threefold W, but in this case W has order n/ν, and V, W are in $(\nu, 1)$ correspondence.

c) The characteristic series is of order zero: in this case the characteristic *system* (of curves) is compounded of a *congruence* (∞^2 system) of index unity or, as it is sometimes expressed, of an involution of curves. Each surface of the system contains a pencil, rational or irrational, of curves of the congruence.

A linear system whose characteristic system and characteristic series are both simple is said to be *completely irreducible.*

4. Characters of a linear system. Suppose, in the first place, that $|A|$ is a linear system of surfaces on V which is effectively free from base points and which has freedom $r \geqq 3$. If we assume further that the characteristic system is simple, then the characteristic curve $\mathfrak{A} = (A^2)$ will in general be irreducible and the characteristic set $\alpha = (A^3)$ non-null; in this case we may compute the effective *arithmetic genus* of A, which is variously denoted by p_A, or $p(A)$, or $[A]$, the effective *curvilinear genus* $\pi_A = \pi(A) = [A^2]$, and the effective *grade* $n_A = n(A) = [A^3]$.

If, however, the characteristic system of $|A|$ is compound, the curve $\mathfrak{A}$ is reducible, in which case the number $\pi(A)$, which is calculated from the formula for the genus of a reducible curve, represents a virtual genus.

In the case where $|A|$ possesses base elements, the situation is more complicated. Suppose, as before, that $r \geqq 3$; then, even if all the base elements (which may be simple or multiple) are assigned, the above numerical characters may correspond to virtual entities — and all of them will certainly do so when A is reducible. If, again, there exist unassigned base elements, with or without assigned base elements, further difficulties will arise. In all such cases we may thus have to distinguish between the virtual and effective characters of $|A|$, including, possibly, the virtual and effective freedoms (for which see III).

We have so far assumed that $r \geqq 3$: the systems for which $r < 3$ are dealt with in II, and for the present we shall suppose that it is possible to define their virtual characters.

It will now be clear that any discussion of the base elements of linear systems must present great difficulties; these are augmented by the fact that little is known concerning the possibility of resolving multiple base points or curves of a given linear system. In particular, it would be desirable to know whether infinitely near multiple base elements of a system can be resolved into distinct base elements.

For these reasons it is obvious that, in computing the characters of a linear system, one usually regards the base elements as virtually non-existent: in most of the theory, as it stands at present, the base elements are effectively non-existent (cf. V, 9).

The concepts of addition and subtraction of linear systems carry over without modification from the theory of surfaces. First, as regards addition: let $|A|$, $|B|$ denote two systems of surfaces*) on V, each virtually free from base points; then the virtual characters of the system $|A + B|$ are given by the formulae (SEVERI [1]):

$$n(A + B) = [A^3] + 3\,[A^2B] + 3\,[AB^2] + [B^3]\,,$$
$$\pi(A + B) = [A^2] + [B^2] + 2\,[AB] + 2\,[A^2B] + 2\,[AB^2] - 3\,,$$
$$p(A + B) = [A] + [B] + [AB]\,.$$

From these formulae we can calculate step by step the characters of the systems $|A + B + C|$, $|A + B + C + D|, \ldots,$ and thence, in particular, those of the system $|hA|$, where h is any positive integer; the latter are as follows:

$$n^{(h)} = h^3 n_A,\ \pi^{(h)} = h^2(\pi_A - 1) + h^2(h - 1)\,n_A + 1\,,$$
$$p^{(h)} = \binom{h}{3}(n_A - 1) + \binom{h}{2}\pi_A + hp_A + \binom{h-1}{3}.$$

Next, we define the difference between two surface systems, exactly as for two curve systems on a surface; we then introduce the concept of a virtual surface as the difference between two effective surfaces, and the null surface as the unique entity $A - A$, where A denotes any surface of V. Using the above formulae, we may proceed to compute the virtual characters of the system $|A - B|$; and we may show that the above formulae for the system $|hA|$ remain valid for negative integral values of h. We may also show that the null surface has virtual grade 0, virtual curvilinear genus 1, and virtual arithmetic genus -1.

5. Jacobian systems. The canonical system. Let $|A|$ be a *general* web, i. e. an ∞^3 system which does not contain ∞^1 reducible members; and suppose that it is effectively free from base points. We then define the Jacobian A_j of the web as the (surface) locus of points which are nodes of surfaces of the system. We may further show that, if $|A|$ is a general system of freedom $r > 3$ (i. e. not containing ∞^{r-2} reducible members), without base points, then the Jacobians of all webs extracted from $|A|$ belong to one and the same linear system, called the Jacobian system $|A_j|$ of $|A|$.

*) Unless the contrary is stated, it is to be assumed in the sequel that the linear systems of surfaces which we consider are free from base points: thus their generic members are irreducible.

It is necessary to restrict the discussion in the first place to general linear systems in order to obtain a determinate Jacobian system: thus, if $|A|$ possessed a *fundamental surface* (having no free intersection with the generic A) or were of freedom less than 3, we should have to modify our procedure in the manner indicated below.

We must now consider the effect of imposing on $|A|$ an ordinary s-fold base point, curve or surface. Since we are here concerned with differential properties, it will suffice to examine a web of surfaces in ordinary space which possesses an ordinary s-fold base point, line or plane; we then find that these elements are respectively $(4s-2)$-fold, $(4s-1)$-fold and $4s$-fold for the Jacobian of the system.

It follows from this that, if $|A|$, $|B|$ are general systems, free from base points, we have the linear equivalences

$$(A+B)_j \equiv A_j + 4B \equiv B_j + 4A \,.$$

These express the fundamental property of the Jacobian. In the first place they can be used to define formally the Jacobians of non-general systems. In the second place they give the relation

$$A_j - 4A \equiv B_j - 4B \,,$$

which shows that, if $|A|$ is general, the system $|K| = |A_j - 4A|$ is invariant. In the case where V is a primal of order n, with ordinary singularities, $|K|$ is evidently obtainable as follows: taking $|A|$ to be the system of prime sections, we see that A_j is the intersection of V with a first polar, residual to the double surface D. Hence the complete system $|K|$, if effective, is cut on V, residually to D, by adjoint primals of order $n-5$; these are called the *canonical adjoints* (§ 2); and we may thus identify the system $|K|$ defined in § 2 with the one just obtained. The number of linearly independent canonical adjoints is called the *geometric genus* P_g of V; when such adjoints do not exist we write $P_g = 0$.

In the case where V is a non-singular threefold of S_r $(r > 4)$ which is the partial intersection of $r-3$ primals of orders $n_1, n_2, \ldots$ whose residual intersection is V', the complete system $|K|$ is cut on V by primals of order $\Sigma n_i - r - 1$ passing through V' (SEVERI [1]).

When the system $|K|$ is virtual, it may happen that the *pluricanonical systems* $|iK|$ are effective for some values of the positive integer i. If the system $|iK|$ is effective, its dimension is written as $P_i - 1$, and P_i is called the i-genus of V; otherwise we take P_i to be zero.

We observe that *the system* $|K|$ *cuts any given surface* A *in a system of curves which is the residual of the characteristic system of* $|A|$ *with respect to the impure canonical system of* A. This follows at once from the fact that $K \equiv (A_j - 3A) - A$, and that, from the theory of surfaces, we know that the curve $(A_j - 3A)A$ is an impure canonical curve of A.

Among various alternative methods of defining the canonical system, the following is perhaps the most important; it was first given in SEVERI (a) for the case of surfaces and then extended (SEVERI [11]) by induction to varieties of any dimension. Consider three pencils $|A|$, $|B|$, $|C|$ of V, each belonging to a general web free from base points; then the locus of a point P such that the surfaces of the three pencils passing through P have a common tangent there is a surface T, which we call the Jacobian of the pencils: we shall show that

$$K \equiv T - 2(A + B + C) .$$

Assuming that the analogous result has been established for surfaces, consider the intersection of T with the general surface A of $|A|$; this consists of the base curve (A^2) of the pencil, and the Jacobian curve $\mathfrak{U}$ on A of the pencils $|(A, B)|$, $|(A, C)|$, which, by our assumption, is expressed by the equivalence

$$\mathfrak{U} \equiv 2[(A, B) + (A, C)] + \mathfrak{K} ,$$

where $\mathfrak{K}$ is an impure canonical curve of A. Thus

$$(T, A) \equiv (A, A) + 2(A, B) + 2(A, C) + \mathfrak{K} .$$

This means that the virtual surface $T - 2(B + C)$ cuts on A a curve equivalent to $(A, A) + \mathfrak{K}$, and hence independent of $|B|$ and $|C|$. If we now substitute for $|B|$ and $|C|$ analogous pencils $|\overline{B}|$ and $|\overline{C}|$, the surface $\overline{T} - 2(\overline{B} + \overline{C})$, where $\overline{T}$ is the Jacobian of $|A|$, $|\overline{B}|$, $|\overline{C}|$, cuts on A a surface equivalent to that cut by $T - 2(B + C)$; hence, by the equivalence criterion (II, 1),

$$\overline{T} - 2(\overline{B} + \overline{C}) \equiv T - 2(B + C) ,$$

so that

$$\overline{T} - 2(A + \overline{B} + \overline{C}) \equiv T - 2(A + B + C) .$$

Thus, if we leave unchanged one of the pencils and alter the other two, the surface $K \equiv T - 2(A + B + C)$ is changed into an equivalent surface; repeating this argument we conclude that K is independent of the particular choice of the pencils. Finally, the identification of $|K|$ with the virtual impure canonical system follows from the equivalence

$$(K, A) \equiv \mathfrak{K} - (A^2) .$$

Another method of obtaining the system $|K|$, likewise due to SEVERI, is as follows. Let $|A|$, $|B|$ denote respectively a pencil and a net of surfaces, each variable in an ∞^3 irreducible system which is free from base points and fundamental varieties; then the locus of contacts of a surface A with a surface B is a surface W, say, which, as we may show, satisfies the equivalence

$$W - 2A - 3B \equiv A_j - 4A .$$

That is to say, $W - 2A - 3B \equiv K$.

Each of the above methods may be extended by induction so as to furnish a definition of the virtual impure canonical hypersurface on any variety V_d $(d > 3)$; and the second method lends itself to a generalisation in which we consider on V_d $t(\geqq 2)$ linear systems $|A^1|$, $|A^2|, \ldots, |A^t|$ of respective dimensions $d_1, d_2, \ldots, d_t$, where $d_1 + d_2 + \cdots + d_t = d$: then, with an analogous definition of the hypersurface W, we may establish the equivalence

$$K \equiv W - \sum_{1}^{t} (d_i + 1) A^i .$$

Returning to the threefolds, we remark that, in certain applications of the theory, a transcendental definition of P_g may prove more convenient. Suppose that V is a primal of S_4, endowed with ordinary singularities, and having the equation, in non-homogeneous coordinates, $f(x, y, z, t) = 0$; then P_g is definable as the number of linearly independent integrals of the form $\int\int\int \frac{\varphi(x, y, z, t)}{f_t} dx\,dy\,dz$, which are everywhere finite on V. Evidently the primal $\varphi(x, y, z, t) = 0$ must be adjoint to V, so that this definition is equivalent to the one previously given.

We have next to examine the behaviour of the system $|K|$ under birational transformation; for simplicity of statement we refer to a non-singular threefold V which is transformed birationally into another non-singular threefold V'. By considering the transform of a Jacobian system it is easily seen that, when the correspondence between V and V' is biregular (i. e. without exceptional elements), the system $|K|$ transforms into the analogous system $|K'|$. A detailed analysis of birational transformations, of varieties of any dimension, which introduce exceptional elements, is given in SEVERI [11]; from this we may conclude that, if the system $|K|$ is effective, all the exceptional surfaces on V are fixed components of $|K|$; they are therefore finite in number. The system residual to these fixed components is called the *pure canonical system* of V; it may be irreducible or reducible with (possibly) fixed parts and base points or curves or both.

It follows from SEVERI's work that the pure canonical system, if effective, is an absolute invariant for birational transformations of V. If, however, $|K|$ is virtual but the system $|iK|$ is effective, the latter contains all the exceptional surfaces as fixed components, so that it again follows that these are finite in number. Thus the number P_i is an absolute invariant of V.

A complete study of the exceptional surfaces has not yet been undertaken, but the following results may be noted. If the neighbourhood of a point of V is transformed to an irreducible non-singular surface E' of V', we call E' *an exceptional surface of the first species*; while, if the neighbourhood of an irreducible non-singular curve of V

is transformed to an irreducible non-singular surface E', we call E' *an exceptional surface of the second species.* It was remarked by PANNELLI [2] that, in the first case, E' is an assigned double component and, in the second, an assigned simple component, of the impure canonical system of V'.

A question which at once presents itself is whether, in the case where V has some plurigenus greater than zero, it is always possible to eliminate the exceptional surfaces by suitable birational transformation. We know that an analogous result holds for surfaces; in fact, irremovable exceptional curves (of the second kind) can exist only on rational or scrollar surfaces. Various examples of threefolds which have $P_g = P_i = 0$, and which are free from exceptional surfaces are known (see V, 2); but it has still to be seen whether or not irremovable exceptional surfaces can occur on threefolds endowed with effective pluricanonical systems. It would also be desirable to undertake a study of reducible exceptional surfaces, but the difficulties confronting such a study appear to be considerable.

6. The arithmetic genus. The irregularities. The *arithmetic genus* P_a of a primal V with ordinary singularities is defined (SEVERI [1]) by the formula

$$P_a = \binom{n-1}{4} - d(n-5)\,,$$

where $d(n-5)$ denotes the virtual postulation of the double surface D for the canonical adjoints. The arithmetic genus of a non-singular threefold in higher space is defined as that of its general projection on S_4.

An expression for P_a in terms of the projective characters of V may be obtained as follows (TODD [2]). Consider the section of D by primals of order m through the triple curve $\mathfrak{C}$ of V (and D); for all sufficiently large values of m such primals cut on D, residually to $\mathfrak{C}$, a linear system $|\mathfrak{C}_1|$ which is complete, non-special and regular, so that its freedom can be inferred from the RIEMANN-ROCH theorem for surfaces; and, at the same time, these primals will have no singularities other than the conic nodes, which they are constrained to have, at the ξ quadruple points of V. The characters of the system $|\mathfrak{C}_1|$ may be determined by familiar correspondence methods; finally, using some formulae of ROTH [1], TODD finds that

$$P_a = \binom{n-1}{4} - \binom{n-4}{2}\alpha_0 + (n-5)(\pi-1) - (p_a+1) + (n-5)t + \xi\,.$$

Here π and p_a denote respectively the sectional genus and arithmetic genus of D, given by the formulae

$$2(\pi-1) = \alpha_1 - 2\alpha_0,\ p_a = \alpha_0 - 1 - \frac{2}{3}\alpha_1 + \frac{1}{6}\alpha_2 + \frac{1}{12}(\beta_2 + \tau)\,.$$

Now, from the expressions for the characters Ω_0, Ω_1, Ω_2 obtained in § 2, we observe that P_a given by the formula

$$2P_a = \Omega_0 - \Omega_1 + \Omega_2 + 4 .$$

This is SEVERI's *relation* (SEVERI [1]), which we have thus established by purely projective methods (ROTH [4, 5, 6]). No other proof of an elementary character is known at present: SEVERI's own demonstration, which employs concepts of birational geometry, is complicated.

While the characters Ω_i are only relative invariants, P_a is an *absolute invariant* for birational transformations of one non-singular threefold into another. An outline of SEVERI's proof of this result will be given in § 10; a different demonstration is due to B. SEGRE [1, 2].

It follows, then, that the number $q_3 = P_g - P_a$ is an absolute invariant of V; this is called the *tridimensional irregularity* of V: as will appear from examples, it may be zero, positive or negative.

Another important absolute invariant of V was discovered by CASTELNUOVO and ENRIQUES [1]; taking V to be a primal with ordinary singularities, they showed by transcendental methods that every surface (termed *generic*) of V which can vary in an ∞^2 linear system with irreducible free characteristic curve has the same irregularity q_2. This number is called the *superficial (bidimensional) irregularity* of V; obviously it may be defined as the irregularity of the generic prime section of V. It is also equal to the number of linearly independent simple integrals of the first kind on V.

It will be seen from examples that a threefold may be bidimensionally or tridimensionally irregular, or both. A threefold for which q_2 and q_3 are both zero is called *completely regular.*

Alternative demonstrations and interpretations of this result have been given in SEVERI [8] and B. SEGRE [2]. SEVERI establishes the existence of the invariant q_2 by reasoning which is partly geometrical and partly dependent on properties of continuous curve systems on a surface*). Also using transcendental methods, B. SEGRE proves that, if C is a generic surface of V, belonging to a net Σ, then q_2 is equal to the number of linearly independent surfaces of the system $|3C + K|$ which pass through the Jacobian curve of Σ. A proof of the preceding result, by the methods of modern algebra, has been given for the case of abstract varieties by MATSUSAKA [1].

The irregularity q_2 has an obvious connection with the existence, on V, of continuous systems of surfaces which are not contained in

*) We may note that ANDREOTTI [1], quoting KODAIRA [1], has given a rigorous proof of the completeness of the characteristic series of any complete continous system of arithmetically effective curves on a surface. An analogous result, for any sufficiently ample continuous system of hypersurfaces on a given variety, has been established by KODAIRA [4], using the theory of harmonic integrals. See also MATSUSAKA [1].

linear systems. Thus, in the first place, if V contains an irrational pencil of surfaces, we must have $q_2 > 0$. And while it is probable that not every complete continuous system of surfaces on V has a complete characteristic system, it has been shown by B. SEGRE [2] that any system $|C|$ cut on V by adjoint primals of sufficiently high order belongs to a continuous system $\{C\}$ endowed with this property; or, in other words, that *the adjoint primals of any sufficiently high order cut on* V *a (complete) linear system the defect of whose characteristic system is equal* to q_2. From this SEGRE has deduced*) that *the deficiency of the characteristic system of any complete linear system, of freedom* $r > 1$, *of surfaces on* V *cannot exceed* q_2.

Suppose now that $|K|$ is the pure canonical system of a given threefold; then, if the characteristic system of $|K|$ is simple, and the freedom of $|K|$ is at least 4, it may be mapped on the prime sections of a threefold W, called a *canonical threefold,* which will be a simple or multiple model, according as the characteristic series of $|K|$ is simple or compound. Examples of canonical threefolds of both types have been given in ROTH [11], where upper limits for P_g are obtained which are analogous to those found by NOETHER and CASTELNUOVO for the geometric genus of a surface (ENRIQUES, a). Thus, if the characteristic series of $|K|$ is complete and the generic characteristic curve K^2 is irreducible, and without fixed points, it follows from CLIFFORD's theorem that $2(P_g - 3) \leqq \Omega_0$, the sign of equality holding if, and only if, K^2 is hyperelliptic, in which case W is a double space. If, further, the characteristic series is simple, it can be shown that $6P_g \leqq 3\Omega_0 + 42$. The canonical threefolds for which the equality sign holds in this relation can be constructed by a simple rule.

These results are generalised in ROTH [12] to varieties V_d of any dimension $d > 3$, the theory of which we now consider briefly. The concepts of the canonical system and its invariant characters extend at once to any variety V_d, where $d > 3$, which is either non-singular or else is a primal with ordinary singularities. In this case we have a set of characters Ω_i $(i = 0, 1, \ldots, d-1)$ representing the grade, curvilinear genus, superficial arithmetic genus, ..., of the virtual impure canonical system $|X_{d-1}|$; these satisfy a set of linear relations, due to ALBANESE [3]; the first of these, namely $2\Omega_1 - 2 = d\Omega_0$, was rediscovered by B. SEGRE [3]. A different set of relations were later obtained by TODD and MAXWELL [1]; in TODD [16] it is shown that the two sets are equivalent.

The arithmetic genus P_a, defined in the first instance for a primal, is an absolute invariant**): as has been shown, in the case $d = 4$,

*) A similar result, for any variety, has been proved by KODAIRA [4], using harmonic integrals.

**) See HIRZEBRUCH [5].

by ALBANESE [2]. SEVERI has conjectured that, for all values of d,

$$\{1 + (-1)^{d-1}\} P_a = \Omega_0 - \Omega_1 + \cdots + (-1)^{d-1} \Omega_{d-1} + (-1)^{d-1} + d.$$

In the case $d = 4$, this conjecture has been confirmed by ALBANESE [1]. The formula has been established for all d by HODGE [1], in the case where there are no multiple base elements.

It is natural to suppose that, for any V_d, there exists a series of numbers $q_2, q_3, \ldots, q_d$ such that q_i ($i = 2, 3, \ldots, d$) is the (constant) i-dimensional irregularity of the generic variety V_i on V_d. No general study of this question has yet been undertaken; all that has been established is the constancy of the irregularity of the generic surface, in the sense already defined, on any V_d ($d > 3$) (SEVERI [8]).

In the next chapter we shall examine other invariant systems than $|X_{d-1}|$, which lead to different invariant characters of V_d.

7. The virtual arithmetic genus. Let V_d be a pure variety (possibly reducible) of S_r ($r > d + 1$) with arbitrary singular varieties of dimension less than d; then it is known (HILBERT [1]) that the postulation of V_d for primals of sufficiently large order l is given by an expression of the form

$$v(l) = \sum_{i=0}^{d} k_i \binom{l+d-i}{d-i},$$

where $k_0, k_1, \ldots, k_d$ are integers depending on V_d and such that $k_0, k_1, \ldots, k_{d-1}$ are the corresponding coefficients in the postulation formula for the general prime section of V_d. We now introduce the numbers p_i defined by the relations

$$p_i = (-1)^i (k_0 + k_1 + \cdots + k_i - 1) \qquad (i = 0, 1, \ldots, d).$$

We then have

$$v(l) = \sum_{i=0}^{d} (-1)^i (p_{i-1} + p_i) \binom{l+d-i}{d-i} \qquad (p_{-1} = 1),$$

while the postulation of the section of V_d by the general S_{r-d+h} given by

$$\varphi(l) = \sum_{i=0}^{h} k_i \binom{l+d-i}{h-i}.$$

The number p_d is called the *virtual arithmetic genus* of V_d (SEVERI [1]). It is clear that $p_0 + 1$ is equal to the order of V_d; we have then to determine the significance of the characters p_i ($i > 0$).

In the case where V_d is irreducible and non-singular, it is known that p_1 and p_2 are respectively the genus of a curve section and the arithmetic genus of a surface section. We shall see (in § 9) that, when $d = 3$, p_3 is equal to the arithmetic genus of V_3 as defined in § 6. In SEVERI [1] it is indicated that an analogous result holds for all d, provided we may assume that every pure variety can be obtained as the limit of a variety free from multiple points. This hypothesis has not yet been justified completely even for curves.

Assuming, then, that, for $d = 3$, $p_3 = P_a$, we are faced with the problem of calculating the invariants of V in terms of the elementary projective characters (§ 1). To this end we project V into a primal W of S_4; then, in order to use the formulae already obtained for W, we must connect the elementary projective characters of V with the non-additive characters of W which we have introduced in § 1. The equations linking one set of characters with the other have been given in ROTH [1, 2]; from these we finally obtain, in the case $r \geqq 6$, the formulae

$$24(P_a - 1) = \mu_3 - 7\mu_2 + 18\mu_1 - 24\mu_0 + \nu_{21} - 3\nu_2\,,$$
$$\Omega_0 = \mu_3 - 12\mu_2 + 48\mu_1 - 64\mu_0 + \nu_3 + 2\nu_{21} - 12\nu_2\,.$$

The values of Ω_1 and Ω_2 may then be deduced from the relations of PANNELLI and SEVERI. In the case $r = 5$, the same formulae hold, provided that ν_3 is replaced by ϱ or ζ.

The above method of calculation is laborious: on the other hand, the formulae obtained are very simple. Now if it were known *a priori* that p_3 (or P_a) and Ω_0 were *enumerative characters*, i. e. functions of the elementary projective characters, it could be shown that they must be linear functions of those characters, and the required results could then be found by using special cases, e. g. complete intersections of primals. With these assumptions the formulae for the invariants of a V_3 and a V_4 have been obtained in ROTH [3] and TODD [1], while, on the same hypothesis, TODD has shown*) that the numbers p_d and P_a are equal, and that P_a is a relative invariant of V_d.

For a different discussion of the arithmetic genus, which is applicable to almost complex manifolds, reference may be made to HIRZEBRUCH [2–4].

8. The adjoint systems. The system $|A'|$ adjoint to a given system $|A|$ of a threefold V is defined by the relation $|A'| = |A + K|$. Evidently $|A'|$ meets A in curves of the virtual impure canonical system of A. From the property of Jacobians (§ 5) follows at once *the fundamental theorem of adjunction:*

$$|(A + B)'| = |A' + B| = |A + B'|\,.$$

If V is a primal of order n, with ordinary singularities, and A is a prime section, the system $|A'|$, if effective, is cut on V by adjoint primals of order $n - 4$. More generally, the system cut by adjoint primals of any order $n - 5 + h$ $(h \geqq 1)$ is obviously adjoint to the system $|hA|$.

*) KODAIRA and SPENCER [1] have established this result by transcendental methods. TODD's hypothesis has subsequently been justified by the work of HIRZEBRUCH [5].

Given the numerical characters n, π, p of any system $|A|$, we may deduce the corresponding characters of $|A'|$ from the fact that $|A'| = |A + K|$. Thus, denoting by ω the virtual linear genus of A, we obtain the following results (SEVERI [1]):

$$n' = 4n - 6\pi + 3\omega + 3 + \Omega_0 ,$$
$$\pi' = 6n - 9\pi + 4\omega + 5 + \Omega_1 ,$$
$$p' = p + 2n - 3\pi + \omega + 3 + \Omega_2 .$$

Hence

$$n' - \pi' + p' = p + \Omega_0 - \Omega_1 + \Omega_2 + 1 ,$$

or, by SEVERI's relation,

$$n' - \pi' + p' = p + 2P_a - 3 .$$

9. The relation $p_3 = P_a$. We now show, with SEVERI [1] that, for any non-singular threefold W, $p_3 + P_a = \Omega_0 - \Omega_1 + \Omega_2 + 4$. It will then follow from SEVERI's relation that $p_3 = P_a$.

I. Let V be the general projection of W (situated in S_r) on S_4; and let $r^{(h)}$ be the freedom of the (complete) system cut on V, residually to the double surface D, by primals of order $l = n - 5 + h$, where n is the order of V. Now the virtual postulation of D for primals of order l satisfies the relation

$$d(l) = d(l-1) + k_0 l + k_1 ,$$

where $k_0 l + k_1$ is the virtual postulation of a prime section of D. Substituting $l = n - 4$, $n - 3$. . . . , in turn, and adding, we have, for a value $l = n - 5 + h$ for and after which the postulation formula is effective,

$$d(l) = d(n-5) + k_0\left[\binom{l+1}{2} - \binom{n-4}{2}\right] + h k_1 .$$

In terms of P_a this becomes

$$d(l) = \binom{n-1}{4} - P_a + k_0\left[\binom{n-4+h}{2} - \binom{n-4}{2}\right] + h k_1 .$$

Hence

$$r^{(h)} = \binom{n-1+h}{4} - \binom{h-1}{4}$$
$$- \binom{n-1}{4} + P_a - k_0\left[\binom{n-4+h}{2} - \binom{n-4}{2}\right] - h k_1 - 1 .$$

Now the prime section A of V has sectional genus $\pi = \binom{n-1}{2} - k_0$, and arithmetic genus $p = \binom{n-1}{3} - k_0(n-4) - k_1$. Hence, using the formulae of § 5, we can calculate the corresponding characters $\pi^{(h)}$

and $p^{(h)}$ of $|hA|$. Inserting the results in the above equation for $r^{(h)}$, we obtain $r^{(h)} = p^{(h)} - 1 + P_a$.

II. We can also calculate $r^{(h)}$ by reasoning from W. Through W we draw $r - 3$ primals, of orders $n_1, n_2, \ldots,$ such that their residual intersection cuts W in an irreducible non-singular surface Φ; then the complete system $|(hB)'|$, where $|B|$ denotes the system of prime sections of W, is cut on W by primals of order $s + h$ passing through Φ, where $s = \Sigma n_i - r - 1$. We suppose the numbers n_i so large that the postulation formula $\varphi(l)$ for Φ is effective from $l = s + h$ onwards; then, if $w(l)$ denotes the postulation formula for W, we have

$$r^{(h)} = w(s + h) - \varphi(s + h) - 1 .$$

Now, if $b(l)$ denotes the postulation formula for B,

$$w(s + 1) = w(s) + b(s + 1) .$$

From this relation we obtain, by addition,

$$w(s + h) = w(s) + \sum_{i=1}^{h} b(s + i) .$$

We next observe that, when the primals of order s cut on W the complete system adjoint to Φ, i. e. the complete canonical system on Φ, the primals of order $s + h$ will cut on Φ the (complete) system adjoint to $|h\mathfrak{C}|$, where $\mathfrak{C}$ is a prime section of Φ: hence, by PICARD's theorem,

$$\varphi(s + h) = \pi_h + \tilde{\omega} ,$$

where π_h is the genus of $h\mathfrak{C}$, and $\tilde{\omega}$ the arithmetic genus of Φ. We thus have

$$r^{(h)} = w(s) - \tilde{\omega} + \sum_{i=1}^{h} b(s + i) - \pi_h - 1 .$$

Now

$$b(s + i) = n \binom{s + i + 1}{2} - (s + i)(\pi - 1) + p + 1 .$$

Also $\mathfrak{C}$ has order $n(s + 2) - 2(\pi - 1)$ and genus $n\binom{s+2}{2} - (s + 1)(\pi - 1) + 1$. Hence $\sum_{i=1}^{h} b(s + i) = n\left[\binom{s + h + 2}{2} - \binom{s+2}{2}\right] - (\pi - 1)\left[\binom{s + h + 1}{2} - \binom{s+1}{2}\right] + hp + h$. Finally, the value of π_h can be found from § 5; then, equating the two expressions for $r^{(h)}$, we obtain $w(s) = P_a + \tilde{\omega}$.

III. Supposing, then, that W has virtual arithmetic genus p_3, and that the postulation formula for W is effective for $l = h$, we see from the formulae of § 8 that the freedom $R^{(h)}$ of the complete system $|hB|$ is given by $R^{(h)} = n^{(h)} - \pi^{(h)} + p^{(h)} - p_3 + 2$. Hence, if the numbers n_i are so large that the formula $w(l)$ holds for $l = s$, we can put $w(s) = R^{(s)} + 1$, so that $P_a + \tilde{\omega} = n^{(s)} - \pi^{(s)} + p^{(s)} - p_3 + 3$.

Again, since the system $|sB|$ is adjoint to Φ, it follows from § 8 that $n^{(s)} - \pi^{(s)} + p^{(s)} = \tilde{\omega} + \Omega_0 - \Omega_1 + \Omega_2 + 1$. Thus, in conclusion, $P_a + p_3 = \Omega_0 - \Omega_1 + \Omega_2 + 4$, whence $p_3 = P_a$.

An analogous result for fourfolds has been established by ALBANESE [1], using a different method.

10. Fundamental property of adjoint systems. Absolute invariance of P_a. Let V be any non-singular threefold, and A any irreducible non-singular surface of V: an important theorem of SEVERI [1] states that *the deficiency of the (canonical) system cut on A by $|A'|$ cannot exceed $q_2 + q_3$, and that there exist surfaces of V for which this maximum is attained.* The proof, which we outline below, depends on two lemmas.

Lemma 1. Let $|B|$, $|C|$ be any irreducible systems of V such that $|(B + C)'|$ cuts a complete system on B: then the deficiency of the system cut on B by $|B'|$ does not exceed the deficiency of the system cut on C by $|C'|$.

Lemma 2. Let A be an irreducible surface on which is given an irreducible system $|\mathfrak{C}|$, free from base points, but not necessarily complete. Let $|\mathfrak{A}|$ be a complete regular system of A which cuts on $\mathfrak{C}$ a complete non-special series containing the characteristic series of $\mathfrak{C}$ and having a non-special series for residual: then the minimal sum $|\mathfrak{A} + \mathfrak{C}|$ is complete and regular.

This follows from the result (CASTELNUOVO, a, p. 431) that the minimal series containing every set of the series $|(\mathfrak{A}\mathfrak{C})|$, and every set $\mathfrak{C}^2$, is complete and non-special.

Let M denote a prime section of V, and consider the system $|hM + A'|$, which cuts on A a system, not necessarily complete, of curves $\mathfrak{C}$. For all sufficiently large values of k, the primals of order k cut on A a complete regular system $|\mathfrak{A}|$ satisfying the conditions of Lemma 2 with regard to $\mathfrak{C}$. Hence the minimal system $|\mathfrak{A} + \mathfrak{C}|$ is complete, i. e. the system cut on A by the system $|(h + k)M + A'|$, which is adjoint to $|(h + k)M + A|$, is complete.

By Lemma 1, the deficiency δ_1 of the system cut on A by $|A'|$ does not exceed the deficiency δ_2 of the system cut by $|((h + k)M)'|$ on $(h + k)M = C$, say. Writing $h + k = l$, we have for the freedom of $|C'|$ the formula $r = p_a^{(l)} - 1 + P_a$, where $p_a^{(l)}$ is the arithmetic genus of lC. Again, since $|C'|$ cuts on C a system of freedom $p_g^{(l)} - 1 - \delta_2$, we have

$$r = p_g^{(l)} - 1 + P_g - \delta_2 .$$

Thus $\delta_2 = (P_g - P_a) + (p_g^{(l)} - p_a^{(l)})$. Supposing, as we may, that lC is generic, we obtain $\delta_2 = q_3 + q_2$. And since $\delta_1 \leqq \delta_2$, it follows that $\delta_1 \leqq q_2 + q_3$. It also follows that *the number $q_2 + q_3$ is non-negative.*

Moreover, since P_g and q_2 are absolute invariants of V, we see that P_a must likewise be an absolute invariant*).

11. Examples. As in the theory of surfaces, much valuable experimental material is to be obtained from the study of particular examples. We consider first what is perhaps the most important of these, namely the threefold which represents the product of a curve and a surface; this was investigated by SEVERI [1], using transcendental methods, but a number of results have been re-obtained geometrically by GAETA [2], who has extended them to the product of any number of given manifolds.

I. Consider the threefold $V = \mathfrak{C} \times C$, where $\mathfrak{C}$ is a curve of genus π, and C a surface, with invariants p_g, p_a, ω, which we assume to be free from exceptional curves of the first kind. We suppose that V is a non-singular model, situated in some higher space; and we wish to determine its virtual canonical system $|K|$. To begin with, it is clear that, if C contained an exceptional curve of the first kind, this would give rise to an exceptional surface on V: conversely, GAETA has conjectured that, if C contains no such curves, then the surface K is pure.

Using the transcendental definition of the geometric genus, we see at once that $P_g = p_g \pi$; this result will be obtained incidentally from GAETA's method of forming the equivalence for K, based on the second alternative definition of $|K|$ described in § 5. Let $|\mathfrak{A}|$ be a general net on C, and $|\alpha|$ a general linear series of freedom 1 on $\mathfrak{C}$, both free from base points; denote by W_1 the curve $P_1 \times \mathfrak{C}$, where P_1 is a point of C, and by W_2 the surface $P_2 \times C$, where P_2 is a point of $\mathfrak{C}$; and consider the surface

$$K \equiv T - 3\mathfrak{A} \times W_1 - 2\,\alpha \times W_2\,,$$

where T is the locus of points at which the surfaces $\mathfrak{A} \times W_1$ and $\alpha \times W_2$ have intersection multiplicity two; by reasoning similar to that of § 5 it may be shown that the surface K in question is a virtual canonical surface of V. Now the intersection of $\mathfrak{A} \times W_1$ and $\alpha \times W_2$ is the curve $\alpha \times \mathfrak{A}$; hence a necessary and sufficient condition that this curve should have a double point at $P = P_1 \times P_2$ is that one at least of the points P_1, P_2 should be double for $\mathfrak{A}$ or α. Since the locus of such double points is the corresponding Jacobian J_1 or J_0, say, while the canonical varieties K_1, K_0 are given by $K_1 \equiv J_1 - 3\mathfrak{A}$, $K_0 \equiv J_0 - 2\,\alpha$, we see that

$$K \equiv K_1 \times W_1 + K_0 \times W_2\,.$$

Since the canonical systems of $\mathfrak{C}$ and C have respective grades $2\pi - 2$ and $\omega - 1$, it follows that $\Omega_0 = 6\,(\pi - 1)\,(\omega - 1)$. We also find that

*) For a different proof of the absolute invariance of p_3, applicable only to arithmetically normal threefolds, see MUHLY and ZARISKI [1].

$\Omega_2 = 3(\pi - 1)(\omega - 1) + 2(\pi - 1)p_a + 2\pi - 3$. Hence, by SEVERI's relation, $P_a = (\pi - 1)p_a + \pi$. The result $q_2 = p_g - p_a + \pi$ is obtained at once by transcendental reasoning.

The geometrical method given above is applied by GAETA to the product of any number of varieties. It may thus be shown that the geometric genus of the product is equal to the product of the geometric genera of the separate factors. Also, using the postulational definition of the arithmetic genus, GAETA proves that, for a variety $V_t = V_{d_1} \times V_{d_2} \times \times V_{d_3} \ldots$, the arithmetic genus p_t is given by the formula

$$p_t + (-1)^t = \Pi\{p_{d_i} + (-1)^{d_i}\}.$$

where p_{d_i} denotes the arithmetic genus of V_{d_i} $(i = 1, 2, \ldots)$.

II. We note here some applications of the above results for threefolds which will be required later. Incidentally, these show that it is possible to have $q_2 = 0$, $q_3 > 0$; or $q_3 = 0$, $q_2 > 0$; or $q_3 < 0$. In the case where C is rational, we have a threefold generated by a pencil, of genus π, of rational surfaces; again, when $\mathfrak{C}$ is rational, the threefold is generated by a congruence, of index 1, of rational curves, the congruence being birationally equivalent to the surface C. In each case the invariants of the threefold are given by the above formulae.

III. By reasoning which is partly geometrical and partly transcendental, SEVERI [1] also obtains the invariants of the threefold which maps the triads of points of an irreducible curve of given genus π; he finds that $P_g = \binom{\pi}{3}$, $P_a = \binom{\pi}{3} - \binom{\pi}{2} + \pi$, $q_2 = \pi$, $\Omega_2 = \binom{\pi}{2} - \pi$.

IV. Consider now a primal V whose singularities are not ordinary; in this case each example must be discussed on its own merits, since no general theory of multiple points, curves and surfaces exists at present. The first step consists in determining the effect of these singularities on the adjoint primals, it being understood that, when V is transformed birationally into a non-singular threefold W, the canonical system of V is to transform into that of W. It is easy to show that an ordinary s-fold point, ordinary s-fold non-singular curve, and ordinary s-fold non-singular surface must have the respective multiplicities $s-3$, $s-2$, $s-1$ on the adjoint primals; and that, if V has order n, the virtual canonical system of V is cut by adjoint primals of order $n-5$. As regards P_a, this is defined by means of the appropriate virtual postulation formula in each case: but in order to give significance to the definition, it is desirable to show that the number so obtained is equal to the arithmetic genus of W.

The postulation formulae required in the above cases are actually known. More generally, the postulation of a multiple curve for primals of any space has been determined, first tentatively in TODD [3] by assuming that it can be calculated from a degenerate form of the curve, namely a set of lines having the appropriate number of intersections, and then rigorously by B. SEGRE [4]. For a non-singular surface of S_4, the postulation formula is given in ROTH [13]; a tentative extension to the case of a surface with a finite number of improper nodes has been given by TODD [3], on the assumption that the surface may degenerate into a set of planes.

V. An important class of primal with a higher type of point singularity is represented by the equation, in non-homogeneous coordinates, $t^2 = f(x, y, z)$, where f is a polynomial of degree $2n$. This primal may be projected from the singular point in question into a double space S_3, whose branch surface F has order $2n$. Assuming that F is non-singular, we may show, exactly as in the theory of double planes (ENRIQUES, a) that the complete canonical system of the primal is mapped on S_3 by the surfaces of order $n-4$, from which it follows that

$$P_g = P_a = \binom{n-1}{3}, \quad \Omega_0 = 2(n-4)^3 .$$

In the case where F acquires an ordinary singularity its influence on the canonical system is readily determined: thus an ordinary $2s$-fold point of F must be $(s-2)$-fold on the images of the canonical surfaces, while an ordinary $2s$-fold non-singular curve of F must be $(s-1)$-fold on these images.

VI. In the last example, if F acquires a number of ordinary singularities, or singularities which are not ordinary, it may well happen that the primal is thereby rendered either superficially or tridimensionally irregular, or both. A study of such double spaces, like the corresponding study of double planes, might yield valuable results.

It will be recalled that the earliest examples of irregular non-scrollar surfaces were constructed by CASTELNUOVO (a) by considering surfaces with singularities at eight associated points. We can, by analogy, construct irregular threefolds with singularities at a set of points which are a complete intersection of primals, or with a multiple curve which is likewise a complete intersection. Such types have been considered by MAXWELL [1] and HALL [1] respectively. HALL has also obtained irregular types which possess composite multiple surfaces; from these examples it appears that the imposition of a multiple surface may actually increase the arithmetic genus of a threefold, in contrast with what happens in the theory of surfaces.

12. Threefolds of linear genus unity. It is a remarkable fact that the majority of non-scrollar surfaces which have been studied in any detail belong to one vast family; those members of the family which possess an effective canonical or pluricanonical system are characterised by the property that their absolute linear genus $p^{(1)}$ is equal to unity; and, by an extension of this concept, it is possible to include the elliptic scrollar surfaces in the scheme. It therefore seems that a useful beginning to the classification of threefolds might be made by examining those types which in a sense to be made precise, have linear genus unity.

Since a surface for which $p^{(1)} = 1$ has a virtual pure canonical system of grade zero, we consider by analogy a threefold V whose virtual pure canonical system $|K|$ has character $\Omega_0 = 0$ (whence $\Omega_1 = 1$, and $\omega = 1$). Supposing, in the first instance, that $|K|$ has no fixed components, we have four main classes of threefold to consider (ROTH [28]):

I. The first class comprises those threefolds for which the pluricanonical systems are all virtual; these are analogous to the elliptic scrollar surfaces.

II. The next class consists of threefolds possessing only isolated canonical or pluricanonical surfaces (possibly of order zero).

III. Next, we have the threefolds whose canonical and pluricanonical systems, when effective, are compounded of a pencil (rational or irrational) of surfaces of zero virtual grade. Here the characteristic curves of these systems have order zero.

IV. In this case the characteristic curves are elliptic, forming a congruence of index unity. Each canonical and pluricanonical surface is generated by a pencil of elliptic curves.

We have now to examine the possibility that $|K|$ has fixed components: here, in contrast with the theory of surfaces, the results so far obtained are less precise. To begin with, we may relax the conditions in III, so as to obtain a system $|K|$ having for fixed components a number of surfaces each of which is a submultiple of K; or, when K is virtual, but iK is effective, we may apply similar considerations to $|iK|$. Such threefolds have an obvious analogy with the surfaces which contain pencils of elliptic curves; and they are dealt with in ROTH [28]. By the same analogy we may construct threefolds of class IV whose canonical systems have fixed components. For the congruence Γ of elliptic curves $\mathfrak{C}$ will in general contain an aggregate (finite or infinite) of s-fold elliptic curves of the form $s\mathfrak{C}_s \equiv \mathfrak{C}$, where the number s takes various values. The curves $\mathfrak{C}_s$ will either be isolated or will generate a number of surfaces B_s; and the isolated curves $\mathfrak{C}_s$ are then $(s-1)$-fold for $|K|$, while the surfaces B_s are $(s-1)$-fold components of $|K|$.

In ROTH [31] the threefolds containing congruences Γ of the simplest types are examined. In the first place we consider a congruence Γ which contains no nodal members and no reducible members other than the curves $s\mathfrak{C}_s$; thus every surface belonging to Γ is elliptic or hyperelliptic: in the second place we allow Γ to contain reducible members but stipulate that the irreducible members shall be birationally equivalent, so as to secure that every surface belonging to Γ shall be paraelliptic (ENRIQUES, a).

We may mention here various threefolds of classes II–IV which will be important in the sequel. In class II we have the PICARD threefold, with canonical surface of order zero, and superficial irregularity 3; this includes as particular case the JACOBI threefold, which maps the triads of points of a curve of genus 3. In class III we have the hyperelliptic threefold, and in class IV the elliptic threefold, each of which is invariant for a continuous group of automorphisms whose trajectories form a pencil of hyperelliptic surfaces or a congruence of elliptic curves, as the case may be (see VI). All these threefolds have arithmetic genus unity; and they include as particular cases the improperly Abelian threefolds which map superficially irregular involutions on a PICARD threefold. Other classes of threefold with linear genus unity are the paraelliptic and parahyperelliptic types which are in a sense analogous to the preceding (ROTH [31]).

Chapter II.

Systems of Equivalence.

1. Introduction. The concept (already employed in the last chapter) of linear equivalence of surfaces on a threefold, or of hypersurfaces V_{d-1} on a V_d, is an immediate extension of the analogous theory for surfaces, and presents no new features; it will, however, be convenient to recall here a useful criterion for the equivalence of surfaces on a threefold, due to SEVERI [5]:

If, on a (non-singular) threefold, two surfaces A, B cut equivalent curves on the irreducible surfaces of a pencil Σ (rational or irrational), then either A and B are equivalent or they differ by surfaces contained partially or totally in Σ. If Σ is rational and contains no reducible members, the second alternative is excluded whenever Σ has a base curve or when A and B have the same order.

This theorem can be applied to the case where Σ is a pencil selected from a congruence Γ, of index 1, of curves on the threefold: if the surfaces A, B cut equivalent curves on every surface belonging to Γ, and do not contain curves of Γ, they must be equivalent.

The more general concept of equivalence, which applies to subvarieties of any dimension on a given manifold, was first developed by SEVERI (for the literature see SEVERI, a) in the case of surfaces; shortly afterwards B. SEGRE [1] made a systematic investigation of equivalence relations on a threefold which not only yielded important new results but suggested far-reaching developments in the theory of varieties of higher dimensions.

We begin by recalling the notion of series of equivalence on a surface which, for the present, we assume to be non-singular and irreducible.

We first define an *elementary series*, or *series of intersection*, as the aggregate of point sets common to curves belonging to two given linear systems (where it is understood that, if the systems have a common component, certain sets of the series must be defined by a limiting process). A general series of equivalence is then defined as an aggregate of virtual sets obtained by addition and subtraction of a finite number of elementary series. Series (of points) and systems (of curves) of equivalence on a threefold are defined in an analogous manner.

The theory of equivalence on a threefold V which, in all that follows, is assumed to be non-singular, requires for its development a knowledge of the corresponding results for curves and surfaces; in particular, the establishment of the invariant series and systems of V rests on the theory of invariants of curves and surfaces. A curve has a single invariant series of equivalence, namely its canonical series. A surface S, which we suppose to be non-singular and irreducible, has three invariant series. To begin with, denoting by $\mathfrak{K}_s$ an impure canonical curve of S, we have the canonical series, a set of which is denoted by $\varphi_s \equiv (\mathfrak{K}_s^2)$. Next, to obtain a second series (of SEVERI), we consider a pencil $|\mathfrak{A}|$ of curves on S, with simple base points; if α, $\varkappa_{\mathfrak{A}}$ and $\delta_{\mathfrak{A}}$ denote respectively the base group, a canonical set and the Jacobian group, a SEVERI set ψ_s will be given by $\psi_s \equiv \delta_{\mathfrak{A}} - \alpha - 2\varkappa_{\mathfrak{A}}$. Lastly, a set ε_s of the ENRIQUES series is given by $\varepsilon_s \equiv \varphi_s + \psi_s$.

Denoting by o_s, i_s, p_s the virtual linear genus, the ZEUTHEN-SEGRE invariant and the arithmetic genus of S, we have

$$[\varphi_s] = o - 1, \quad [\psi_s] = i + 4, \quad [\varepsilon_s] = 12(p_s + 1)\,.$$

From these results follows NOETHER's relation

$$o_s + i_s = 12p_s + 9\,.$$

So far we have dealt with curves and surfaces which were assumed to be effective and irreducible. We have now to extend the above concepts to reducible and also to virtual entities. The first group of results, concerning the intersection of surfaces, are obtained exactly as in the theory of curve systems on a surface, and present no new features: thus, given any surface on V, possibly reducible, or isolated, or virtual, we can in all cases define for it a virtual characteristic system and a virtual characteristic series, thereby giving meaning to the symbols (S^2), (S^3) in all cases.

Next, we can define the canonical series $|\varkappa_{\mathfrak{C}}|$ on a curve $\mathfrak{C}$ which is effective but reducible, though virtually free from multiple points: we shall take this to be the virtual series cut on $\mathfrak{C}$ by the system $|\mathfrak{C}'|$ adjoint to $\mathfrak{C}$ on any irreducible surface A containing $\mathfrak{C}$. We easily show that this definition is independent of A; in fact it leads to the well known formula for the genus of a reducible curve. We then extend the definition to the case where $\mathfrak{C}$ is virtual.

2. Covariant point sets. In all subsequent developments the following result (SEVERI, a) is fundamental: *equivalence on any subvariety of V implies equivalence on V.*

Let $\mathfrak{C}$ be any virtual curve of V of the form $\mathfrak{C} = \sum_1^r s_i \mathfrak{C}_i$, and let A be any effective surface of V, not containing any component $\mathfrak{C}_i$. In this case the virtual set $\Sigma s_i(\mathfrak{C}_i A)$, which we denote by $(\mathfrak{C}A)$ or $(A\mathfrak{C})$, is defined; the number of points in the set is written as $[\mathfrak{C}A]$ or $[A\mathfrak{C}]$. Then, from a relation of the form $A \equiv A_1 + A_2$ follows $(\mathfrak{C}A) \equiv (\mathfrak{C}A_1) + (\mathfrak{C}A_2)$; again, from $A \equiv A^*$ we have (on $\mathfrak{C}$ and hence on V) $(\mathfrak{C}A) \equiv (\mathfrak{C}A^*)$. We can now define the series of equivalence (on $\mathfrak{C}$ and in V) characterised by the set $(\mathfrak{C}A)$, even when A contains a component of $\mathfrak{C}$, or is virtual. For let A_1, A_2 be two surfaces meeting $\mathfrak{C}$ in a finite number of points and such that $A \equiv A_1 - A_2$; then it suffices to assume that $(\mathfrak{C}A) \equiv (\mathfrak{C}A_1) - (\mathfrak{C}A_2)$. This definition agrees with the preceding in the case where A is effective and meets $\mathfrak{C}$ in a finite number of points. Moreover, it depends only on $\mathfrak{C}$ and A; for if we take two surfaces A_1^*, A_2^* meeting $\mathfrak{C}$ in a finite number of points, and such that $A \equiv A_1^* - A_2^*$, then $A_1 - A_2 \equiv A_1^* - A_2^*$, whence $A_1 + A_2^* \equiv A_2 + A_1^*$, and so $(\mathfrak{C}A_1) - (\mathfrak{C}A_2) \equiv (\mathfrak{C}A_1^*) - (\mathfrak{C}A_2^*)$.

In particular, if $\mathfrak{C}$ is the complete intersection of the surfaces B, C, so that $\mathfrak{C} = (BC)$, we have thus a meaning for the symbol (ABC) in all cases.

Again, denoting by X a virtual, impure canonical surface of V, we obtain a virtual covariant series of equivalence of sets $\chi_{\mathfrak{C}} \equiv (\mathfrak{C}X)$; the order $[\chi_{\mathfrak{C}}] = x_{\mathfrak{C}}$, say, of the series is called the *immersion character* or *canonical number* of $\mathfrak{C}$.

We next define the *virtual characteristic series* of a curve $\mathfrak{C}$ on a surface of V, even when the surface does not contain $\mathfrak{C}$ or is virtual. Suppose that $\mathfrak{C}$ is effective (possibly reducible), and that A is an effective surface passing simply through $\mathfrak{C}$; then a characteristic set $(\mathfrak{C}^2)_A$ is defined by taking any other surface B through $\mathfrak{C}$ meeting A residually in a curve $\mathfrak{D}$, and writing $(\mathfrak{C}^2)_A \equiv (\mathfrak{C}B) - (\mathfrak{C}\mathfrak{D})_A$.

If two surfaces A, A^* through $\mathfrak{C}$ are equivalent and free from multiple points on $\mathfrak{C}$, then $(\mathfrak{C}^2)_A \equiv (\mathfrak{C}^2)_{A^*}$; for we have, with an obvious meaning for $\mathfrak{D}^*$, $(\mathfrak{C}^2)_A \equiv (\mathfrak{C}B) - (\mathfrak{C}\mathfrak{D})_A \equiv (\mathfrak{C}B) - (\mathfrak{C}\mathfrak{D}^*)_B \equiv (\mathfrak{C}^2)_{A^*}$.

Now suppose that A is a surface virtually free from multiple points on $\mathfrak{C}$, which is equivalent to a surface consisting of two parts, of which one, A_1, say, passes through $\mathfrak{C}$ and is free from multiple points on $\mathfrak{C}$, while the other, $A_2 \equiv A - A_1$, meets $\mathfrak{C}$ in a finite number of points. Then the former curve $\mathfrak{D}$ breaks up into $\mathfrak{D}_1$, intersection of A_1, B residual to $\mathfrak{C}$, and $\mathfrak{D}_2 = (A_2B)$. Thus

$$(\mathfrak{C}\mathfrak{D}_2)_B = (\mathfrak{C}A_2), \ (\mathfrak{C}^2)_{A_1} \equiv (\mathfrak{C}B) - (\mathfrak{C}\mathfrak{D}_1)_B ,$$

whence

$$(\mathfrak{C}^2)_A \equiv (\mathfrak{C}B) - (\mathfrak{C}\mathfrak{D}_1)_B - (\mathfrak{C}\mathfrak{D}_2)_B \equiv (\mathfrak{C}^2)_{A_1} - (\mathfrak{C}A_2)$$
$$\equiv (\mathfrak{C}^2)_{A_1} - \{(\mathfrak{C}A) - (\mathfrak{C}A_1)\} .$$

Hence $(\mathfrak{C}^2)_A + (\mathfrak{C}A) \equiv (\mathfrak{C}^2)_{A_1} + (\mathfrak{C}A_1)$. This establishes the required result in the case where the surface $A_2 = A - A_1$ is effective, but this restriction can easily be removed by the procedure already considered above. We thus see that, if A is any effective surface passing through $\mathfrak{C}$ and virtually free from multiple points on $\mathfrak{C}$, the set $(\mathfrak{C}^2)_0 = (\mathfrak{C}^2)_A + (\mathfrak{C}A)$ defines a series of equivalence which depends only on $\mathfrak{C}$. This result can be extended to the case where A has a finite number of multiple points on $\mathfrak{C}$. In all cases we can now assign a meaning to the symbol $(\mathfrak{C}^2)_A$, even when A does not pass through $\mathfrak{C}$ or is virtual: this is called a *characteristic set* of $\mathfrak{C}$ on A, and $[\mathfrak{C}^2]_A$ is termed the *virtual grade* of $\mathfrak{C}$ on A.

We now establish the relation $(\mathfrak{C}^2)_0 \equiv \varkappa_{\mathfrak{C}} - \chi_{\mathfrak{C}}$, where it is assumed that $\mathfrak{C}$ is virtually free from multiple points.

Let A be an effective surface through $\mathfrak{C}$ which is virtually free from multiple points; an adjoint surface $A' \equiv A + X$ meets A in a virtual impure canonical curve $\mathfrak{K}_A$ which cuts $\mathfrak{C}$ in a set residual to the characteristic series with respect to the canonical series, i. e.

$$(\mathfrak{C}A') \equiv \varkappa_{\mathfrak{C}} - (\mathfrak{C}^2)_A .$$

The required result follows from the fact that $(\mathfrak{C}A') \equiv (\mathfrak{C}A) + \chi_{\mathfrak{C}}$.

The concepts introduced above, and the formulae deriving from them, lead to a remarkable simplification of the problems of intersection and equivalence for surfaces on a threefold (including S_3). In B. SEGRE [1] they are applied to the case of surfaces with a simple or multiple common curve; while in BARKER [1] they are used in problems where the surfaces have contact along a common curve. It is obvious that these methods extend immediately to varieties of any dimension: in TODD [4] they are applied to problems concerning surfaces on a V_4 having simple common curves, and in ARCHBOLD [1] to the case of common multiple curves (See also BARKER [2]).

In B. SEGRE [7] the general problem of intersections, both normal and abnormal, of subvarieties on a manifold of any dimension, is solved by a method of great power and simplicity; for this see BALDASSARRI (a).

3. Invariant series and systems. It follows from the preceding results that, if the curve $\mathfrak{C}$ varies in a system of equivalence and the surface A varies in a linear system, then the sets $(\mathfrak{C}A)$, and also the sets $(\mathfrak{C}^2)_A$, belong to series of equivalence. Important particular cases of such sets are $(\chi_{\mathfrak{C}})$ and $(\varkappa_{\mathfrak{C}})$.

Consider any system $|A|$ virtually free from base points; associated with it we have the characteristic system of equivalence $\mathfrak{A} \equiv (A^2)$, and also the system of equivalence defined by (AX). On every surface A we have the system $\mathfrak{K}_A \equiv (AA') \equiv (A^2) + (AX)$; in particular, taking $A = X$, we have the invariant systems defined by $\mathfrak{X} = (X^2)$, and $\mathfrak{K}_X = 2\,\mathfrak{X}$.

The characteristic sets $\alpha = (A^3)$ also vary in a series of equivalence; hence, taking $A = X$, we obtain an *invariant series* $\xi = (X^3)$.

Now we have seen in § 2 that $\varkappa_{\mathfrak{A}} \equiv 2\,(A^3) + (A^2X) = 2\alpha + \chi_{\mathfrak{A}}$. Taking $A = X$, we have $\chi_{\mathfrak{X}} \equiv \xi$, $\varkappa_{\mathfrak{X}} \equiv 3\xi$; thus the numerical content of this result is, in the usual notation, $3\,\Omega_0 = 2\,\Omega_1 - 2$, which is PANNELLI's first relation (I, 2).

We obtain another covariant series of equivalence defined by the sets $(A\mathfrak{X}) \equiv (AX^2)$. Now

$$(A'^2) \equiv \mathfrak{X} + (A^2) + 2(AX), \quad (A\mathfrak{X}) \equiv (AA'^2) - (A^3) - 2(A^2X),$$

from which it follows that $(A\mathfrak{X}) \equiv (\mathfrak{K}_A^2)_A - \alpha - 2\chi_{\mathfrak{A}}$. This last relation gives

$$(A\mathfrak{X}) \equiv \varphi_A + 3\alpha - 2\varkappa_{\mathfrak{A}}\,,$$

which shows that the sets $\varphi_A \equiv (A\mathfrak{X}) - 3\alpha + 2\varkappa_{\mathfrak{A}}$ vary in a series of equivalence. Taking $A = X$, we obtain $\varphi_X \equiv 4\,\xi$, which is the functional counterpart of PANNELLI's second relation, $\omega = 4\,\Omega_0 + 1$. Thus, in the present theory, PANNELLI's equations appear as numerical aspects of equivalence relations which are themselves merely special cases of general functional results; we shall see shortly that SEVERI's equation is capable of a similar interpretation.

Writing the above equivalence in the form $\varphi_A \equiv (A^3) + 2(A^2X) + (AX^2)$, we observe that it enables us to define the canonical series $|\varphi_A|$ in all cases, even when A is reducible or virtual.

A second *invariant series* of equivalence is obtained as follows: consider a general pencil $|A|$ of surfaces, of which $\mathfrak{A} = (A^2)$ is the base curve and δ_A the Jacobian set; then it may be shown that the set

$$\zeta \equiv \delta_A - \varkappa_{\mathfrak{A}} - 2\psi_A$$

varies in a series of equivalence (the SEVERI series) which is independent of $|A|$. The order of the series so obtained is $I - 6$, where I denotes the ZEUTHEN-SEGRE invariant of V. The method of proof is exactly similar to that employed in establishing the existence of the ZEUTHEN-SEGRE invariant of a surface: we consider the locus of contacts of two rational pencils of surfaces generically situated.

Evidently the above relation may be used to define the Jacobian set of any pencil $|A|$ whatever.

It is worth noting that, just as in the theory of surfaces, the invariant I may be calculated by using an irrational pencil of surfaces, provided the general surface of the pencil is non-singular; thus if $\{A\}$ is a pencil of genus $\varrho > 0$, satisfying this condition, we find that

$$I - 6 = \delta + 2(\varrho - 1)(i + 4).$$

The next stage in the development of the theory consists in determining the contribution made to the value of I by a surface of a given pencil (rational or irrational) which has singularities of prescribed type or is reducible with, possibly, multiple components. An important general result of this kind will be given in VI, 4.

While the system $|X|$ and the series $|\zeta|$ present no novel features, being strictly analogous to entities already encountered in the theory of surfaces, the *invariant system of curves* which we now introduce typifies an entirely new development.

Let $|S|$ be a net of surfaces whose base points are simple and finite in number; and let $\mathfrak{J}_s$ be the Jacobian curve of the net. We shall show (B. SEGRE [1]) that the virtual curve

$$\mathfrak{Y} \equiv \mathfrak{J}_s - 6(S^2) - 3(SX)$$

varies in a series of equivalence which is independent of $|S|$. We here follow SEGRE's exposition, referring later to an alternative method of procedure.

Consider two nets Θ and Θ_1 of surfaces S and S_1 whose base points are simple (not points of contact), forming two sets α and α_1 with no points in common. Let $\mathfrak{R}$ denote the curve of contact of the nets, i. e. the locus of points at which a surface of Θ and a surface of Θ_1 have a common tangent line. It is clear that $\mathfrak{R}$ passes simply through α and α_1, while the Jacobian curves $\mathfrak{J}_s$, $\mathfrak{J}_{s_1}$ of Θ and Θ_1 do not contain any of these points. Through any point P not belonging to α or α_1 there passes a single characteristic curve $\mathfrak{A}$ of Θ and a single characteristic curve $\mathfrak{A}_1$ of Θ_1. If P lies on $\mathfrak{J}_s$ or $\mathfrak{J}_{s_1}$, $\mathfrak{A}$ or $\mathfrak{A}_1$ (as the case may be) has a double point at P, and conversely; the points P of $\mathfrak{R}$ which do not also lie on $\mathfrak{J}_s$ or $\mathfrak{J}_{s_1}$ are characterised by the fact that the corresponding curves $\mathfrak{A}$, $\mathfrak{A}_1$ are tangent there; and a point P which lies on $\mathfrak{R}$ and $\mathfrak{J}_s$ is double for $\mathfrak{A}$, simple for $\mathfrak{A}_1$, and the three tangents at P to these curves are coplanar.

Let Φ denote the pencil of curves $\mathfrak{A}$ on the general surface S, and let Ψ_1, Ψ_1^* denote two pencils of surfaces of Θ_1, of which $\mathfrak{A}_1$, $\mathfrak{A}_1^*$ are the base curves and S_1 the common surface; these pencils cut S in two pencils Φ_1, Φ_1^* having in common the curve $\mathfrak{D} = (SS_1)$. Now consider on S the curves $\mathfrak{T}$, $\mathfrak{T}^*$ loci of points where a curve of Φ touches a curve of Φ_1 or of Φ^*; as is well known (SEVERI, a),

$$\mathfrak{T} \equiv \mathfrak{T}^* \equiv 2\mathfrak{A} + 2\mathfrak{D} + \mathfrak{K}_s.$$

We may therefore write

$$\mathfrak{T} \equiv \mathfrak{T}^* \equiv 3\,(S^2) + 2\,(SS_1) + (SX)\,.$$

It is clear that $\mathfrak{T}$, $\mathfrak{T}^*$ pass through the base group α of Φ, touching at each such point the curve of Φ_1 or the curve of Φ_1^* which contains it, so that $\mathfrak{T}$, $\mathfrak{T}^*$ have simple intersections there. Also $\mathfrak{T}$ passes through the base group $(\mathfrak{A}_1 S)$, which does not lie on $\mathfrak{T}^*$.

Through any point P distinct from $(\mathfrak{A}_1 S)$ and $(\mathfrak{A}_1^* S)$ there passes a single curve of each of the pencils Φ, Φ_1, Φ_1^*. If P is common to $\mathfrak{T}$, $\mathfrak{T}^*$, it is double for the curve $\mathfrak{A}$ of Φ which contains it, i. e. it lies on $\mathfrak{J}_s$; or it is a contact of this curve with each of the curves of Φ_1, Φ_1^* passing through P; and conversely. In this last case these two curves can coincide with $\mathfrak{D}$, and then P is double for the $g^1_{[\mathfrak{A}\mathfrak{D}]}$ cut by Φ on $\mathfrak{D}$; otherwise P lies on $\mathfrak{R}$; and conversely.

Denoting by λ a Jacobian set of $g^1_{[\mathfrak{A}\mathfrak{D}]}$, we thus have

$$(\mathfrak{T}\mathfrak{T}^*) \equiv (\mathfrak{J}_s S) + \lambda + (\mathfrak{R}S)\,.$$

Now let S vary in a general pencil Ψ of Θ: then the sets $(\mathfrak{J}_s S)$, $(\mathfrak{R}S)$ generate respectively the curves $\mathfrak{J}_s$, $\mathfrak{R}$, while the set λ varies on the Jacobian curve $\mathfrak{L}$ of the net cut by Θ on S_1. Denoting by T, T^* the surfaces described by $\mathfrak{T}$, $\mathfrak{T}^*$, we therefore have

$$(TT^*) \equiv \mathfrak{R} + \mathfrak{J}_s + \mathfrak{L}\,.$$

Again the Jacobian curve $\mathfrak{L}$ is given by the equivalence

$$\mathfrak{L} \equiv 3\mathfrak{D} + \mathfrak{K}_{s_1}\,.$$

Substituting for $\mathfrak{D}$ and $\mathfrak{K}_{s_1}$, we obtain

$$\mathfrak{L} \equiv 3(SS_1) + (S_1^2) + (S_1 X)\,.$$

Since, given the generality of the pencils Ψ, Ψ_1, Ψ_1^*, neither T nor T^* passes through the base curve of Ψ, we have

$$(ST) \equiv (ST^*) \equiv \mathfrak{T} \equiv \mathfrak{T}^* \equiv (S, 3S + 2S_1 + X)\,.$$

Hence, by the equivalence criterion of § 1,

$$T \equiv T^* \equiv 3S + 2S_1 + X\,.$$

Substituting for $\mathfrak{L}$, T, T^*, and simplifying, we obtain

$$\mathfrak{R}_s + \mathfrak{J}_s \equiv 9(S^2) + 3(S_1^2) + 6(SX) + 3(S_1X) + 9(SS_1) + (X^2)\,.$$

Now, by interchanging the nets Θ, Θ_1, we obtain a relation of the same form, in which S and S_1 are interchanged; equating the two equivalences for $\mathfrak{R}$, and re-arranging, we find that

$$\mathfrak{J}_s - 6(S^2) - 3(SX) \equiv \mathfrak{J}_{s_1} - 6(S_1^2) - 3(S_1X)\,.$$

This result demonstrates the invariance of the curve $\mathfrak{Y}$ defined above. We call $\mathfrak{Y}$ a *virtual canonical curve* of V. Incidentally, the last equivalence enables us to define the Jacobian system of equivalence $|\mathfrak{J}_s|$ in regard to any surface S of V.

Supposing that S satisfies the conditions of generality already laid down, we easily obtain the equivalence

$$(\mathfrak{Y}S) \equiv \psi_S - (S^3) - (S^2X) .$$

This enables us to define the SEVERI series of any surface whatever. And, taking $S = X$, we at once have

$$(\mathfrak{Y}X) \equiv \psi_X - 2\xi = \eta , \text{ say.}$$

We thus obtain a third *invariant series* of equivalence $|\eta|$ on V; substituting for $[\psi_X]$ and $[\xi]$, we find that the order $[\eta]$ of this series (which we call the SEGRE series) is given by

$$[\eta] = 12(\Omega_0 - \Omega_1 + \Omega_2 + 2) .$$

Hence, by SEVERI's relation, $[\eta] = 24(P_a - 1)$.

In point of fact, SEVERI's relation is most easily established by using the above equivalence for η, together with TODD's formula for P_a (ROTH [14]). If the latter formula is to be avoided, an alternative definition of P_a is required, except when the number can be found from first principles, as in an example worked by B. SEGRE [1].

B. SEGRE proves that, if V is transformed birationally into a non-singular threefold V' in such a way as to introduce a single exceptional surface of either species, then $[\eta] = [\eta']$, i. e. for any such transformation P_a is an absolute invariant.

4. Further developments. The six invariant entities $|X|$, $|\mathfrak{X}|$, $|\mathfrak{Y}|$, $|\xi|$, $|\eta|$, $|\zeta|$ defined above form a complete set in the sense that a wide class of enumerative and functional problems can be solved in terms of these alone. For example, B. SEGRE [1] uses the present methods to obtain equivalences for the invariant series of any surface of the form $c_1S_1 + c_2S_2$, where c_1, c_2 are integers, in terms of the invariant series of S_1 and S_2. In the same work SEGRE finds the covariant systems of one or two nets of surfaces on V, of two or more pencils, and also of linear systems of freedom three or four, thereby establishing many interesting relations between the entities in question. One of the most striking of these is the following: given two pencils $|A|$, $|B|$ of general character, generically situated, the number of pairs A, B which have stationary contact with one another is $48(P_a + p_A + p_B + g_{AB})$, where p_A, p_B, g_{AB} denote respectively the arithmetic genera of A, B and the genus of the curve AB. This result, and others of a similar character, can be used to furnish alternative definitions of P_a.

In B. SEGRE [2] the relations between the invariant series of equivalence of two non-singular threefolds in multiple correspondence are studied, and formulae connecting the numerical invariants, analogous to the well known equations (SEVERI, a) between the invariants of two surfaces in multiple correspondence, are deduced from them. The correspondences considered have ordinary branch (and coincidence) elements, with, possibly, ordinary fundamental elements; in the particular case where there are no branch elements – including, of course, that of birational transformation – SEGRE re-obtains certain results previously found by NOETHER and PANNELLI; the appropriate references are given in SEGRE's paper.

5. Extension to varieties of higher dimension. The work of generalising the results of §§ 2, 3 to varieties of any dimension has been carried out by TODD and EGER. To begin with, the existence of invariant systems and series of equivalence on any non-singular V_d was established by TODD [5], subject to a certain hypothesis; contemporaneously, the same results were obtained by EGER, using transcendental methods, though a complete account (EGER [1]) of his work was not published till later. However, the partial account which first appeared suggested to TODD a fresh exposition (TODD [6]) of geometrical character, free from the previous hypothesis; this is essentially a generalisation of the process by which the SEVERI series (or ZEUTHEN-SEGRE invariant) is established for V_d – it has already been used to obtain the canonical system $|K|$ of a threefold in I, 5.

The invariant systems $\{X_h\}$ $(h = 0, 1, \ldots, d-1)$ of V_d, which are called the *canonical systems* of the variety, are obtained by an induction argument on d and h, the result of which is most succinctly stated in terms of the linear operators K_n, due to EGER, whose effect on any subvariety V_t of V_d is as follows:

$$K_n[V_t] = X_n[V_t]\ (t > n); \quad K_n[V_t] = V_t\ (t = n); \quad K_n[V_t] = 0\ (t < n)\,.$$

We consider now $h+1$ general pencils $|S_i|$ of hypersurfaces on V_d, and define their Jacobian $J_h[S_0, S_1, \ldots, S_h]$ as follows: when $h = 0$, the variety is simply the Jacobian set of $|S_0|$, while, when $h > 0$, it consists of the points P such that the $h+1$ tangent $[d-1]$s to the members of the pencils which pass through P have a common $[d-h]$. It may then be shown that the (possibly virtual) variety X_h defined by

$$X_h \equiv J_h[S_0, S_1, \ldots, S_h] - K_h\left[\prod_0^h (1 + S_i)^2 - 1\right],$$

varies in a system of equivalence $\{X_h\}$ which is independent of the pencils $|S_i|$.

It may further be shown that the systems $\{X_h\}$ possess the *property of adjunction* expressed by the relation

$$X_{h-1}[S] \equiv S.\ X_h[S] + X_h[V_d] \qquad \text{(on } S\text{)}.$$

We now define the *canonical invariants* of V_d as the intersection numbers of the systems $\{X_h\}$, where repetition of any system is allowed; there are in all $p(d)$ such invariants, where $p(d)$ denotes the number of partitions of d. The numerical interpretation of this work is given in TODD [1]; there it is shown that every canonical invariant is a homogeneous linear function, with constant coefficients, of the elementary projective characters of V_d; and explicit formulae for the invariants, in the cases $d = 3, 4$ are obtained. It is also shown that, on the assumption that the arithmetic genus p_d is an enumerative character, the number $p_d + (-1)^d$ is a homogeneous linear function, with constant coefficients, of the elementary projective characters.

B. SEGRE's work on covariant systems is the starting point for another series of investigations (TODD [7]) concerning intersection and immersion problems for subvarieties of V_d. Finally, SEGRE's examination of the behaviour of the invariant series of V_3 under birational transformation has led to a study (TODD [8, 9, 10]), with similar motives, of the birational transformations of a V_d which possesses either a fundamental point, curve or surface (see also B. SEGRE [8]).

A novel and greatly simplified exposition of the whole theory is given in B. SEGRE [7]; this stresses the concept, due to TODD [7], of the *ring of equivalence* which is formed by subvarieties of V_d, in which the algebraic operation of multiplication corresponds to the geometrical operation of intersection. As already remarked, this work leads to a remarkable unified treatment of problems of intersection. Again, by first introducing covariant successions of equivalence and then deducing the canonical systems from them, it provides a method of obtaining the latter which is independent of the notion of Jacobian. In B. SEGRE [8] a purely topological definition of the canonical systems is given: for the details see BALDASSARRI (a). An interesting application of SEGRE's methods will be found in VESENTINI [1,2]; for a brief exposition of these and other associated methods and results we may refer to the report by B. SEGRE (b).

One important result of B. SEGRE [7], to which we wish to draw attention here, actually concerns the definition of $X_h(V_d)$ by means of Jacobians. Thus, let S be a *general* hypersurface of V_d i. e. one which, for every h such that $0 \leqq h \leqq d-1$, is variable in a linear system endowed with a well defined pure Jacobian variety $J_h(S)$. With this connotation, let $S_1, S_2, \ldots, S_r$ $(r > d-h)$ be any r general hypersurfaces on V_d; then it may be shown that

$$X_h(V_d) = \Sigma J_h(S_i) - \Sigma J_h(S_i + S_j) + \cdots + (-1)^{r-1} J_h(S_1 + S_2 + \cdots + S_r) .$$

From this we may deduce equivalences for the canonical varieties of any variety V_d which is in biregular $(n, 1)$ correspondence with a variety V_d^*. Suppose that the coincidence locus on V_d is of the form $\Sigma(s-1)\, C_{d-1}^{(s)}$, where the numbers s may assume various values, all of them divisors of d, and where $C_{d-1}^{(s)}$ denotes an $(s-1)$-fold component of the coincidence locus which is non-singular and which has no intersections with any other component; then it may be proved that

$$X_h(V_d) \equiv \bar{X}_h(V_d^*) + \Sigma(s-1)\, X_h(C_{d-1}^{(s)}) ,$$

where $\bar{X}_h(V_d^*)$ denotes the transform of $X_h(V_d^*)$. This result (ROTH [36]) generalises a relation which, in the case $h = d - 1$, is classical.

Chapter III.

Systems of Surfaces.

1. The RIEMANN-ROCH theorem. We consider in this section the problem of determining the freedom of the complete linear system characterised by a given non-singular surface on a non-singular three-fold V. If C is such a surface, with virtual characters n, π, p, we define the *virtual freedom* d of the system $|C|$ by the formula $d = n - \pi + p - P_a + 2$, where P_a denotes the arithmetic genus of V. In the case where C is non-special, with effective freedom $r = d$, we say that $|C|$ is *regular*. It is known that there exist regular systems on V; thus (SEVERI [1]) the surfaces adjoint to any sufficiently high multiple of a linear system which is free from fundamental surfaces form a regular system.

A first step towards obtaining an inequality for r in cases where $|C|$ is not regular was taken by SEVERI [1], following the method originally employed by ENRIQUES in establishing a limited form of the RIEMANN-ROCH theorem for surfaces. Suppose that C is adjoint to an irreducible non-singular surface A, so that we may write $C = A'$; and let p_g, p_a denote the genera of A. Now (I, 10) $|A'|$ cuts on A a linear system of freedom $p_g - 1 - \delta$, where $\delta \leqq q_2 + q_3$; also the freedom r of $|A'|$ is given by the equation $r - (p_g - 1 - \delta) - 1 = P_g - 1$, where P_g is the geometric genus of V. If we further suppose that A is generic, in the sense of I, 6, we may put $q_2 = p_g - p_a$, in which case we have

$$r \geqq p_g - 1 + P_g - (p_g - p_a) - (P_a - P_g) ,$$

whence

$$r \geqq p_a - 1 + P_a .$$

Thus, by I, 8, $r \geqq n' - \pi' + p' - P_a + 2$, which is the required inequality, expressed in terms of the characters of $|A'|$. In terms of the characters of $|C|$ this becomes: *If $|C|$ is adjoint to a generic non-singular surface*, then $r \geqq n - \pi + p - P_a + 2$.

We pass now to the form of the RIEMANN-ROCH theorem for more general systems of surfaces; this is due to B. SEGRE [2], the proof being on the lines of CASTELNUOVO's demonstration of the analogous result in the theory of surfaces. An essential preliminary is the theorem (I, 6) that the deficiency of the characteristic system of any complete linear system (at least ∞^2) of irreducible surfaces cannot exceed q_2; this is proved by projecting V generically into a primal W, and showing that adjoint primals of sufficiently high order cut on W a linear system the defect of whose characteristic system is precisely q_2, and then applying the following lemma:

If the system $|C| = |A'|$ is regular, and is adjoint to a surface A belonging to a continuous system which is not an irrational pencil; and if the system $|D|$ is at least ∞^1, irreducible, and such that $|A + D|$ is irreducible, then $|C + D|$ is regular, and cuts on the generic D a complete, regular and non-special system.

From this result, the proof of which presents no difficulty, we may deduce that the deficiency of the characteristic system of $|D|$ cannot exceed q_2.

We now show that, *if $|C|$ is a complete linear system, irreducible and infinite, then*

$$r \geqq (n - \pi + p - P_a + 2) + i - (q_2 + q_3 + \omega)\,,$$

where i is the index of speciality of $|C|$, and ω is the defect of the system cut on C by the canonical system of V.

For, if j denotes the index of speciality of the characteristic system of $|C|$, the freedom ϱ of the complete system satisfies the inequality

$$\varrho \geqq n - \pi + p - j + 1\,,$$

by the RIEMANN-ROCH theorem for surfaces. Again, since, as we have remarked, $\varrho \leqq r - 1 + q_2$, it follows that

$$r \geqq n - \pi + p + 2 - q_2 - j\,.$$

To evaluate j, we observe that the system residual to $|(C^2)|$ with respect to the canonical system of C belongs to the system cut on C by the canonical system $|X|$ of V, and that the latter system has freedom $P_g - 1 - i$; we therefore have $j = P_g - i + \omega$, whence the above result.

In certain cases it is possible to evaluate ω; thus, when $P_g = 0$, we have $\omega = 0$ (and, of course, $j = i = 0$); and when X is of order zero, $P_g = 1$, $j = 1$, $i = \omega = 0$.

Using the theory of harmonic integrals, KODAIRA [2] has extended SEVERI's form of the RIEMANN-ROCH theorem to more general surface systems. He considers a system $|C| = |X + S|$, where S is a surface (possibly reducible) of V which is endowed with ordinary singularities, and obtains an equation for r in terms of transcendental and geometrical invariants of S. KODAIRA has also verified SEVERI's conjecture that the number of linearly independent double integrals of the first kind on V is $q_2 + q_3$.

2. Applications and extensions. We consider now some of the more important deductions from the preceding results.

I. Let $|A|$ be a system of generic surfaces on V which is free from base points and which has assigned characters n, π, p; then, by SEVERI's form of the RIEMANN-ROCH theorem, the system $|hA + A'|$, which is adjoint to the system $|(h+1)A|$, has an effective freedom ϱ_h given by

$$\varrho_h = \varrho_h' + \varepsilon_h \qquad (\varepsilon_h \geqq 0)\,,$$

where ϱ_h' denotes the virtual freedom of the system; the latter can at once be written down by using the formulae of I, 4.

Now $|hA + A'|$ cuts on A a system of curves which is adjoint to the curves hA^2 and hence, by PICARD's theorem, regular; the freedom r_h of the system is therefore given by

$$r_h = n\binom{h}{2} + h(\pi - 1) + p - \delta_h \quad (\delta_h \geqq 0) .$$

Again, we have the relation $r_h = \varrho_h - \varrho_{h-1} - 1$; substituting for r_h, ϱ_h and ϱ_{h-1}, we find, after reduction, that

$$\varepsilon_h + \delta_h = \varepsilon_{h-1} .$$

This relation is valid for all $h \geqq 1$; for $h = 0$, however, we have

$$\varepsilon_0 + \delta_0 = q_2 + q_3 .$$

Putting $h = 1, 2, \ldots, l$, and adding, we obtain $\varepsilon_l + \sum_0^l \delta_h = q_2 + q_3$. Hence (SEVERI [1]): *for all sufficiently large values of h, the system adjoint to $|(h+1)A|$ cuts a complete system on A.* Moreover, *if V is completely regular, this system, for all $h \geqq 0$, is complete.*

This result, which generalises the theorem of I, 10, may be used to obtain a different generalisation of the latter theorem. We now know that, when V is *completely regular,* both the systems $|A'|$ and $|A + A'|$ if effective, cut complete systems on A; then, from the relation $(A + hA') + (hA + A') = (h+1)(A + A')$, we infer that, *if V is completely regular, the systems $|A + hA'|$ $(h \geqq 1)$, if effective, cut complete systems on A.*

II. The RIEMANN-ROCH theorem may be used to find upper limits to P_g in cases where X is a generic surface. In such cases X is certainly irreducible: we further suppose that it is non-singular. On X the adjoints $X' = 2X$ cut canonical curves, so that the characteristic system $|\mathfrak{X}|$ of $|X|$ on X is self-residual to the canonical system of X. And since $i_X = 1$, it follows that $j_{\mathfrak{X}} = P_g + \omega_X - 1$, while the freedom of $|\mathfrak{X}|$ is $j_{\mathfrak{X}} - 1$.

Let k be the freedom of the series cut on $\mathfrak{X}$ by the canonical system $|2\mathfrak{X}|$ of X. Since X is generic, the latter system has freedom $\Omega_2 + q_2 - 1$, while the residual system $|\mathfrak{X}|$ has freedom $j_{\mathfrak{X}} - 1$; whence

$$(\Omega_2 + q_2 - 1) - (k + 1) = j_{\mathfrak{X}} - 1 .$$

Consider now the characteristic series of $|\mathfrak{X}|$ on $\mathfrak{X}$; this has freedom $j_{\mathfrak{X}} - 2$; and since any two sets of this series are cut on $\mathfrak{X}$ by a curve of $|2\mathfrak{X}|$, it follows that they belong to the above series of freedom k. We thus obtain the inequality $2(j_{\mathfrak{X}} - 2) \leqq k$, so that, from the previous equation, $3(j_{\mathfrak{X}} - 1) \leqq \Omega_2 + q_2$. Substituting for $j_{\mathfrak{X}}$, we have $3(P_g + \omega_X - 2) \leqq \Omega_2 + q_2$. And since $\omega_X \geqq 0$, we finally obtain $3P_g \leqq \Omega_2 - q_2 - 6$. This result is due to B. SEGRE [2], who has found other inequalities of a similar type. New limits for P_g have been obtained by JONGMANS [1].

Using the RIEMANN-ROCH theorem in conjunction with certain enumerative results of B. SEGRE [1], SEGRE has obtained a formula for the number of moduli of a threefold which does not admit automorphisms; this invokes an unproved hypothesis concerning continuous systems of threefolds analogous to that used in calculating the number of moduli of a surface (see ZARISKI, a).

III. Turning now to varieties of higher dimension, we remark that the definitions of the virtual dimension of a linear system of hypersurfaces, and hence of a regular system, on any V_d, are laid down by obvious analogy with the case $d = 3$. Using the theory of harmonic integrals, KODAIRA [4] has established an exact form of the RIEMANN-ROCH theorem for any sufficiently ample system on V_d; he has also obtained a formula*) for the alternating sum of the integrals of the first kind analogous to the one mentioned above. In HIRZEBRUCH [1] the RIEMANN-ROCH theorem is deduced from the theory of analytic stacks, or faisceaux. Another deduction from this work is the result, generalising PICARD's theorem, that adjoint systems of hypersurfaces are regular.

MARONI [1] has shown how to construct complete linear systems of hypersurfaces on any V_d generated by ∞^1 spaces $[d-1]$, and has then established SEVERI's conjectural form of the RIEMANN-ROCH theorem for such a variety (cf. § 1); the method employed is on the lines of earlier work by the same author, in which he constructs complete linear systems of curves on any surface which maps the product of two given curves. This last construction has received its greatest possible generalisation in GAETA [2], where analogous results are obtained for the variety which maps the product of any number of given varieties; from these follow interesting applications to postulation theory, including (as already remarked) a formula for the arithmetic genus of the product variety in terms of the arithmetic genera of its factors.

3. Varieties which contain only complete intersections. An important question which frequently presents itself in the theory is that of determining whether a given threefold contains only complete intersections with primals of the ambient space. We consider first those results which have been obtained by classical methods. To begin with, it has long been known (see BAKER, a) that non-singular quadrics of any dimension exceeding 2, and also quadrics with only a certain degree of specialisation, contain only complete intersections. Next, it has been shown by SEVERI [4] that any non-singular threefold of S_4 contains only complete intersections, from which it follows that an analogous property holds for a non-singular primal of any dimension. SEVERI [10] has extended this result to primals of any dimension $r \geqq 4$ which contain fewer than ∞^{r-3} multiple points.

*) $P_a = g_d - g_{d-1} + \cdots + (-1)^{d-1} g_1$, where g_i is the number of linearly independent i-ple integrals of the first kind. This was conjectured by SEVERI [1].

Another contribution to the problem is the determination, by Fano [4], of the nodal cubic primals of S_4 which can contain partial intersections; it is shown that such primals must have at least 6 nodes, unless 4 of them are coplanar, in which case the plane of the nodes lies on the primal. A simplified proof of this result, based on equivalence criteria, has been given by Archbold [2].

A large class of varieties which contain only complete intersections consists of the Grassmannians and their linear sections of sufficiently high dimensions. A classical result of Severi states that the variety which maps, in the accustomed manner, the $[k]$s of $[n]$, contains only complete intersections; Roth [22] has shown that an analogous property holds for every general linear section of the Grassmannian down to and including the variety which maps the complete intersection of $n - k$ linear $[k]$-complexes.

Severi's result concerning non-singular primals is a particular case of the following theorem: an irreducible non-singular V_d $(d > 2)$ which is the complete intersection of $r - d$ primals of S_r contains only complete intersections. The first attempt to prove this was made by Fano [6], who begins with a surface which is a complete intersection of $r - 2$ primals, and then reasons to a threefold, and so on: but his work is based on the assumption (first justified by Hodge) that a double integral of the first kind on a surface cannot be without periods. The theorem has been proved by Lefschetz [1], using transcendental-topological methods. Lefschetz also proves that a non-singular surface which is a complete intersection of $r - 2$ primals, and which is normal in S_r $(r > 3)$ *in general* contains only complete intersections, i. e. provided the surface is a generic member of the linear system determined by it on a threefold containing it. Lefschetz obtains an analogous result for surfaces of order exceeding 3 in ordinary space, thereby making precise an incomplete theorem due to Noether.

4. Theory of the base. The existence of a finite base for algebraic curves on an algebraic surface was first established by Severi, using transcendental methods; later a topological proof, originating with Lefschetz and simplified by Albanese, was obtained (for the details see Albanese [4]). From this the existence of a base for hypersurfaces on any non-singular V_d (or primal with ordinary singularities) readily follows. An arithmetic proof of the theorem, valid for varieties defined over a field of any characteristic, has been given by Néron [1].

An important consequence of the above theorem is the existence, on V_d, of a finite number $\sigma - 1$ of algebraic zero divisors with respect to algebraic equivalence of hypersurfaces. The number σ is called the *divisor* of V_d; unlike the base number, which is only a relative invariant, *σ is an absolute invariant for birational transformation.*

A useful remark, due to LEFSCHETZ [1], is that, if V_d has base number ϱ and divisor σ, the corresponding characters ϱ', σ' of the prime section of V_d satisfy the inequalities $\varrho' \geqq \varrho$, $\sigma' \geqq \sigma$. LEFSCHETZ (b) has further shown that $\sigma' = \sigma$.

FANO [14] has considered the case of two threefolds V, V' which contain only complete intersections and which are in *regular* birational correspondence: that is, the system of prime sections of V (or V') is mapped on V' (or V) by a system whose base points and base curves are all distinct. FANO shows that the number of base points and base curves on V is equal to the corresponding number on V'. He also shows that any fundamental curves of the second species in the representation must be rational, and obtains a simple relation between the ZEUTHEN-SEGRE invariants of V and V'.

For future reference we recall here some important classes of surface which have divisor unity. In the first place there are the *rational and scrollar surfaces*; and, more generally, any surface which is the product of two curves. In the second place there are the surfaces, regular or irregular, with *an effective (pure) canonical curve of order zero*: the PICARD surface is a case in point. On the other hand, the simplest examples of surfaces with divisor greater than unity*) are afforded by surfaces with a virtual canonical curve of order zero, e. g. the ENRIQUES sextic surface (ENRIQUES, a), for which $\sigma = 2$, and certain elliptic surfaces of geometric genus zero. GAETA [1] has studied a simply-infinite class of surfaces with divisor greater than unity; these have characters $p_g = p_a = 0$, $p^{(1)} = 1$, $\sigma = 2^r$ $(r = 1, 2, \ldots)$ and include the ENRIQUES surface as a special case.

With regard to varieties of higher dimension we may remark that, since any linear space has divisor unity, it follows that *any birational manifold likewise has divisor unity.*

The determination of ϱ and σ is usually very difficult, even for surfaces; the only general results available are for product varieties. Thus, for the variety $V_d = V_{d_1} \times V_{d_2} \times V_{d_3} \ldots$, LEFSCHETZ [1] gives the formulae $\varrho = \Sigma \varrho_i + \Sigma \lambda_{ij}$, $\sigma = \sigma_1 \sigma_2 \sigma_3 \ldots$,
where ϱ_i, σ_i denote respectively the base number and divisor of the factor V_{d_i}, and where λ_{ij} is the simultaneity index of V_{d_i} and V_{d_j}. A detailed proof of the first of these results has been given by SCOTT [1] in the case where $d = 3$, $d_1 = 1$, $d_2 = 2$, and this demonstration could be extended to the general case. A different proof of LEFSCHETZ's formula, which uses the theory of harmonic integrals, is due to IGUSA [1].

The theory of the base has interesting connections with the arithmetic theory of forms, as SEVERI first pointed out for the particular case of surfaces. Consider, for example, a threefold V, on which a base for

*) Numerous examples have been constructed by GODEAUX; see GODEAUX [2].

surfaces is $(A, B, C, \ldots)$, so that any surface on V can be represented by the symbol $xA + yB + zC \ldots$. The virtual grade of this surface is then given by the fundamental form $f(x, y, z, \ldots) = [A^3]\,x^3 + 3\,[A^2B]x^2y \ldots$, a homogeneous cubic polynomial in $x, y, z \ldots$. Suppose that V possesses a birational self-transformation which carries a surface $xA + yB + zC \ldots$ into a surface $x_1A + y_1B + z_1C \ldots$; this will be accompanied by an automorphism, in the integral domain, of $f(x, y, z, \ldots)$, induced by a linear substitution of modulus ± 1. However, it will not in general be the case that every (or indeed any) such automorphism, other than identity, will give rise to a birational self-transformation of V; for such a transformation must carry effective surfaces into effective surfaces. We are thus led to distinguish between *effective* and *virtual* automorphisms of the fundamental form. Again, it may well happen that V possesses birational self-transformations which are not arithmetically ascertainable, i. e. not reflected in automorphisms of $f(x, y, z, \ldots)$. Examples of these phenomena are given in ROTH [15], where the corresponding theory for fourfolds is also illustrated.

The problem of the base for subvarieties V_k of any dimension k on V_d has been solved by SEVERI [7], using topological methods, subject to the hypothesis that, on V_d, arithmetical equivalence of subvarieties implies algebraic equivalence. SEVERI has also established the duality law that the base number for varieties V_k is equal to that of varieties V_{d-k}.

In the case where V_d is either a Grassmannian or a SEGRE variety, the problem has been solved from first principles, independently of the above result. Thus, on a Grassmannian, a base for subvarieties of any given dimension is provided by the SCHUBERT varieties of that dimension (see SEVERI, c; HODGE and PEDOE, a). The various bases for submanifolds on a SEGRE variety have been determined by BENEDICTY [1].

5. Applications of the PICARD variety. Just as in the theory of surfaces an important part is played by those surfaces which contain an irrational pencil of curves, so, in the theory of threefolds, we are led to consider those types which possess an irrational pencil of surfaces; however, by reason of the duality between curves and surfaces on a threefold, equal importance attaches to threefolds which contain irregular congruences (of index unity) of curves. And, generally, we are led to examine varieties V_d which contain systems, of index unity, of varieties V_k $(1 \leq k \leq d-1)$; it is customary in this theory to call all such systems *congruences**); the genus of the pencil $\{V_k\}$, when $k = d-1$, or the superficial irregularity of the system $\{V_k\}$, when $k < d-1$, as the case may be, is termed the *irregularity* of $\{V_k\}$.

*) SEVERI has suggested the term *involution*; but it is convenient to reserve this for systems of point-sets on a variety (see VI, 3).

As was first noticed by CASTELNUOVO (a) in the case $d=2$, the properties of V_d are intimately connected with those of the PICARD varieties. These varieties will be discussed in VI: for present purposes it suffices to remark that a PICARD variety V_p is a non-singular manifold which is endowed with a completely transitive permutable continuous group of ∞^p birational self-transformations, or automorphisms, and that the possession of such a group characterises V_p; also that V_p has a pure canonical hypersurface of order zero. From the latter property it may be deduced (ENRIQUES, a) that any non-exceptional surface on V_p must have geometric genus greater than zero; thence, by induction, we may establish an analogous result for submanifolds of any dimension $k>2$ on V_p.

A first deduction from this result is the theorem (SEVERI [6]): *Any variety of geometric genus zero and superficial irregularity $q>0$ contains a congruence of irregularity q.* To prove this, we consider, on the variety V_d, a system $\{C\}$ of ∞^1 disequivalent hypersurfaces, of index $i>1$, and we denote by $C_i = \Sigma C$, the sum of the i hypersurfaces issuing from the same point of V_d. Now the varieties C_i cannot all be equivalent, for in that case the C's would also be equivalent: moreover, it can be shown that if the C_i's were all disequivalent, or if each were equivalent to a finite number of varieties of the system, then V_d would have geometric genus greater than zero. Hence the system $\{C_i\}$ consists of ∞^r subsystems $(1 \leqq r \leqq d-1)$, the varieties in each subsystem being equivalent; and from this result we may deduce the existence of the congruence in question.

A second theorem, due to COMESSATTI [1], which generalises a result of CASTELNUOVO (a), states that, *if the characters P_g, q of V_d satisfy the inequality $P_g \leqq d(q-d)$, then V_d contains an irregular congruence of varieties.* More precisely, in the case $d=3$, if there is a congruence of curves, its irregularity is at least $q-P_g$ $(\geqq 3)$, while if there is a pencil of surfaces, its genus is at least $q-P_g-1$ $(\geqq 2)$. COMESSATTI's demonstration, like CASTELNUOVO's, is transcendental, but, as D'ORGEVAL [1] has observed in the case $d=2$, a geometrical proof may be given. Following CASTELNUOVO, we consider on V_d a continuous system $\{C\}$ of hypersurfaces consisting of ∞^q disequivalent linear systems: such systems can always be found on V_d. The image of such a system is a PICARD variety V_q which we assume to be non-singular; and, if V_d does not contain a congruence, V_d is mapped, simply or multiply, by a variety W_d (which we suppose is non-singular) on V_q. From the correspondence between V_d and W_d, to which we apply the considerations mentioned below, it readily follows that $P_g > d\,(q-d)$.

Another result due to COMESSATTI [1], also established by transcendental methods, is as follows: if the characters q_2 (>0), q_3 of a threefold V_3 satisfy the inequality $q_3 \leqq q_2 - 4$, then V_3 contains a pencil of surfaces which, except when $q_2=1$ (in which case the pencil is elliptic), has genus 2 at least.

Returning now to the case $d = 2$, we observe that an irregular non-scrollar surface for which $p_g = 0$ necessarily has arithmetic genus $p_a = -1$ (ENRIQUES, a); moreover, it can be shown that, in addition to the irrational (elliptic) pencil which the surface contains, there is a second pencil, which is rational and free from base points. COMESSATTI [2] has shown that the surfaces for which $p_g \geqq 2(p_a + 2)$, and which, by CASTELNUOVO's theorem, contain an irrational pencil, contain a second pencil without base points. Beginning with a system of partial differential equations, COMESSATTI transforms the problem into one of differential line geometry in higher space, and thus arrives at a classification of the surfaces in question. The discussion has been completed in several particulars by NOLLET [1].

ROSENBLATT [1] has applied analogous methods to the threefolds for which $P_g \leqq 3(q - 3)$, but has obtained only partial results.

We have already had occasion to mention the representation of a variety V_d on a multiple W_d; it will be convenient to state at this point some results, geometrical and transcendental, concerning this representation, which will be required later. Suppose then that V_d is mapped on the n-fold variety W_d; in general there will be a branch hypersurface on W_d, possibly reducible, with components of different multiplicities; and, corresponding to this, a coincidence hypersurface on V_d, possibly reducible, with components of different multiplicities. Assuming, for the sake of simplicity, that there are no fundamental elements in the representation, we may show that, *on V_d, a virtual impure canonical hypersurface is equivalent to the transform of a virtual impure canonical hypersurface of W_d, together with the coincidence hypersurface*: this result is proved exactly as in the case $d = 2$. Again, we may show that *the superficial irregularity of V_d is at least equal to that of W_d*. Finally, *if W_d contains a congruence $\{W_k\}$ then V_d must contain a corresponding congruence $\{V_k\}$*. Here various possibilities may arise: thus the involution I_n of sets of points of V_d corresponding to the points of W_d may be simple or compound, and the variety which maps W_k may or may not belong to I_n; in all cases, however, by virtue of the fact that the system $\{W_k\}$ has index unity, the corresponding system on V_d must either consist of, or be compounded of, a congruence $\{V_k\}$. And the irregularity of $\{V_k\}$ is at least equal to that of $\{W_k\}$: this last general result is due to SEVERI [9] – the case $d = 2$ is classical.

Chapter IV.

Criteria of Rationality.

1. Introduction. We consider a projective model V_d of an irreducible algebraic variety, defined in some space S_r $(r > d)$ by means of a set of polynomial equations, the field of rationality of whose coefficients we shall denote by K. If, while remaining in K, we can find a rational parametric representation for V_d, we shall say that V_d is *unirational in* K; we thereby obtain a mapping in K of V_d on an involution I_n of S_d, of a certain order $n \geqq 1$. If, in particular, $n = 1$, it is possible to reverse the parametric equations of V_d, so as to express each of the d essential parameters as a rational function of the coordinates of the general point of V_d: in that case we shall say that V_d is *birational in* K; and the representation will be termed unirational or birational, as the case may be.

In general, however, it will happen that this representation is obtainable (if at all) only in some extension K' of K: in that case we shall say that V_d is unirational (or birational) in K'. It should be noted that these concepts are invariant for birational transformations of V_d which are defined in K and free from exceptional elements.

We begin by recalling the results for $d = 1, 2$, which are classical (and, as will appear, quite untypical). When $d = 1$, we have LÜROTH's theorem, according to which any unirational curve is also birational. A similar result holds*) for $d = 2$ (CASTELNUOVO, a); this is conveniently expressed in the form: *all plane involutions are rational.* By virtue of these theorems we may speak of a *rational* curve or surface whenever we are not concerned with the particular extension K' of K in which its birational representation is obtainable. We shall see later that there exist *irrational* (i. e. non-birational) involutions in any space S_d ($d \geqq 3$), so that *a variety V_d ($d \geqq 3$) which is unirational in K' is not in general birational in any extension of K'.*

We illustrate these ideas by examples which will prove useful in the sequel. To begin with, any *monoid* (primal of order n with a single $(n-1)$-ple point) is obviously birational in K; hence, in particular, the VERONESE surface is birational in K; for its general projection on S_3 is a monoid (STEINER surface). Again, a large class of varieties V_d, including the quadrics, is birational in the extension of K obtained by adjoining the irrationality on which depends the determination of the general point P of V_d: in that case we say that V_d is birational in $K(P)$; when V_d is non-singular it is often the case that P may be any point whatever of V_d. As a third example, consider the DEL PEZZO variety V_d^4, complete intersection of two quadrics of S_{d+2} ($d \geqq 2$); this projects birationally from any one of its lines l into a space S_d: we therefore say that V_d^4 is birational in $K(l)$. And, more generally, the concept of a variety V_d which is unirational or birational in $K(S_h)$, where S_h is some linear space on V_d, is of considerable importance in the theory.

Finally, we observe that a variety V_d which is unirational and representable on an involution $I_{n'}$, in an extension K', may be represent-

*) In general, however, only in some extension of K: see B. SEGRE [10].

able on an involution $I_{n''}$, where $n'' < n'$, in some other extension K''; in particular, we may have $n'' = 1$, in which case V_d is birational in K''. For example, a non-singular cubic surface V_2^3 of S_3 is representable on an involution I_6 in $K(P)$, and is representable on an involution I_2 in $K(l)$ (§ 3), while it projects birationally on to a plane by means of transversals to a pair of skew lines of V_2^3.

2. The standard forms of NOETHER and ENRIQUES. An essential first step in the development of criteria of rationality for threefolds and higher varieties is the birational transformation, in K, of a rational curve or surface to one or other of a certain number of standard forms. First, let V_1^n be a non-singular rational curve in higher space, so that its general projection W_1^n on to a plane will be a curve of order n, with $\frac{1}{2}(n-1)(n-2)$ ordinary nodes. The adjoint curves of order $n-2$ cut on W_1^n a linear series g_{n-2}^{n-2}, determinable in K, which, if $n > 3$, can be mapped on the prime sections of a non-singular normal curve V_1^{n-2}. By repetition of this process, we can transform V_1^n birationally in K either to a twisted cubic or a conic, according as n is odd or even; and in the former case we can further transform V_1^n birationally to a line, for the twisted cubic projects into a monoid. Hence, *any non-singular rational curve of order n can be transformed birationally, in K, to a line or a conic, according as n is odd or even* (NOETHER [1]).

ENRIQUES [1] has effected an analogous reduction of the rational surfaces to one of four standard forms. Let F be a non-singular rational surface of order n and sectional genus $\pi > 1$, so that its general projection on S_3 will be a surface G, of order n and sectional genus π, endowed with ordinary singularities. The linear system cut on G by the adjoint surfaces of order $n-3$ is determinable in K; assuming that the prime sections of F are mapped on a plane by the system $|\mathfrak{C}|$ of curves of order m, we see that the linear system in question will be mapped by the system $|\mathfrak{C}'|$ of curves of order $m-3$, adjoint to $|\mathfrak{C}|$. If now the system $|\mathfrak{C}'|$ can map the prime sections of an irreducible surface F' (which is certainly non-singular), we proceed to consider the analogous system $|\mathfrak{C}''|$ adjoint to $|\mathfrak{C}'|$; and so on. In any case the series of curve systems $|\mathfrak{C}|, |\mathfrak{C}'|, |\mathfrak{C}''|, \ldots$, and with it the series of surfaces $F, F', F'', \ldots$ must terminate after a finite number of stages. It may be shown (ENRIQUES*) [2]) that the last surface of the series has rational, elliptic or hyperelliptic curve sections, or is representable on a double plane with a branch curve which is either a non-singular plane quartic or a sextic with two consecutive triple points (i. e. a double quadric cone with sextic of genus 4 as branch curve). These results may be summarised as follows:

*) Actually the result is due to CASTELNUOVO; it is given in an appendix to this paper.

The non-singular rational surfaces may be classified in four families, according as the process of successive adjunction, applied to their prime sections, leads to a birational representation in K on

1. *a surface with rational curve sections;*
2. *a surface with elliptic curve sections and of order* $n \geqq 2$, *which, for* $n > 2$, *is non-singular;*
3. *a surface with hyperelliptic curve sections, containing a rational pencil of conics;*
4. *a double quadric cone with sextic branch curve.*

It may be added that, if a simple model of 4. is required, we may take it to be a surface F^9 mapped on the plane by novenic curves having 8 triple base points. A simple model of 2., $n = 2$, is provided by a surface F^8 mapped by sextics with 7 double base points.

Using these results, ENRIQUES proceeds to determine the types of irrationality (or extensions K' of K) required to map the surfaces on the plane. For the applications that we have in view, only a selection of his results will be needed, supplemented by some considerations of unirationality which do not enter into his work. Thus we may leave aside the surfaces of the first family, and pass to those of the second family, which are either DEL PEZZO surfaces of order n $(3 \leqq n \leqq 9)$ or the double plane with quartic branch curve $(n = 2)$.

I. $n = 2$: this type is dealt with in § 8, where it is shown that the surface is unirational in $K(P)$.

II. $n = 3$: this surface projects from a point of itself into the previous type. We shall show, however, in § 3 that the surface is unirational in $K(P)$ and representable on an involution I_6.

III. $n = 4$: from a point P of itself the surface projects into a cubic surface, on which we are given a line p, image of P; hence, by § 3, the DEL PEZZO quartic surface is representable in $K(P)$ on an involution I_2.

IV. $n = 5$: the DEL PEZZO quintic possesses a single apparent triple point, which means that we can determine, in K, a pair of points P, Q of the surface. Projecting the surface, in its normal space S_5, from P, Q, we obtain a cubic surface on which we are given a pair of skew lines p, q, images of P, Q; and this surface may be mapped birationally on a plane by drawing transversals to p and q. Thus the DEL PEZZO quintic surface is birational in K.

V. It follows from the last result that the DEL PEZZO sextic surface is birational in $K(P)$, since it projects from P into a DEL PEZZO quintic.

VI. $n = 7$: from the plane representation of this surface, which is by cubics $\mathfrak{C}^3(O_1 O_2)$, we see that it contains three lines, one of which is incident to the other two; from the latter, which is determinable in K, the surface projects into a VERONESE surface, which means that it is birational in K.

VII. $n = 8$: there are two species, the first mapped by cubics $\mathfrak{C}^3(O)$, the second by quartics $\mathfrak{C}^4(O_1^2 O_2^2)$. The first contains a line, determinable in K, from which it evidently projects into a rational quintic scroll, birational in K, while the second*) projects from a tangent plane into a surface of VERONESE type, since its plane representation is by quartics $\mathfrak{C}^4(O_1^2 O_2^2 O_3^2)$, and so is birational in $K(P)$.

VIII. $n = 9$: since the novenic type projects from a point P of itself into an octavic of the first species, it follows that the surface is birational in $K(P)$.

In conclusion, then, *the* DEL PEZZO *surfaces of order n are*

a) *birational* in K *when* $n = 5, 7, 8$ *(first species)*;

b) *birational in* $K(P)$ *when* $n = 6, 8$ *(second species)*, 9;

c) *unirational in* $K(P)$ *when* $n = 3, 4$.

Also, *the double plane with non-singular quartic branch curve is unirational in* $K(P)$.

With regard to the surfaces of the third and fourth families, all that we need note at present is that those of the third family contain a rational pencil of conics, determinable in K, while those of the fourth family contain a rational pencil of elliptic curves, likewise determinable in K.

3. Varieties which are unirational in $K(S_h)$. In the previous section we have encountered various examples of surfaces which are either unirational or birational in the extension $K' = K(P)$; this is a particular case of an extension which has proved to be of considerable importance in the theory, for it frequently happens that a given variety V_d is unirational or birational in $K(S_h)$ where S_h $(h \geqq 0)$ is the generic member of a system of spaces S_h lying on V_d. Before going on to questions of a more general character, we shall therefore examine some of the results which derive from this concept. We begin by noting the following simple criterion (ROTH [37]):

Given on V_d *a birational* ∞^{d-k} *system of varieties* V_k, *of index* $\nu \geqq 1$, *such that* V_k *is unirational in* $K(S_h)$ $(h \geqq 0)$, *if it is possible to determine rationally, on the generic* V_k, *a space* S_h, *then* V_d *is unirational. If, further,* $\nu = 1$, *and* V_k *is birational in* $K(S_h)$, *then* V_d *is birational.*

As a first application we consider the non-singular cubic primal V_d^3 $(d \geqq 2)$, for which we introduce the extension $K(P)$, where P is the general point of V_d^3. The general tangent plane to V_d^3 at P meets V_d^3 in an irreducible cubic with a node at P; and this is birational in $K(P)$. Again, the tangent primes to V_d^3 at points of the cubic cut V_d^3 in a birational system, of index 6, of monoids, each of which is birational in $K(P)$. It follows that V_d^3 is unirational in $K(P)$ and, more precisely, representable in $K(P)$ upon an involution I_6.

*) In ENRIQUES [1] it is stated that this surface is birational in K; the above modification is given in ENRIQUES [7].

Next, denoting by l the generic line on V_d^3, we may consider the system of tangent primes at points of l; we thus obtain a birational system, of index 2, of monoids: hence V_d^3 is unirational in $K(l)$ and representable in $K(l)$ upon an involution I_2.

This last result has a useful application to the DEL PEZZO variety V_d^4 $(d \geqq 2)$, intersection of two non-singular quadric primals, generically situated, in S_{d+2}. Such a variety is obviously birational in $K(l)$, since it projects birationally from a line l of itself into a space S_d. But it is also representable in $K(P)$ upon an involution I_2; for it projects from P into a primal V_d^3 – no longer non-singular – to which the preceding considerations apply.

The fact that the non-singular cubic primal V_d^3 $(d \geqq 3)$ is unirational was known to NOETHER; a far more remarkable result, due to MORIN [7], states that *the general primal of any given order is unirational provided it lies in a space of sufficiently high dimension.* The first step towards this theorem was taken in MORIN [1], where it is shown that – in a convenient extension K' of K – the general quartic primal of S_r is unirational, and representable on I_6, provided that $r \geqq 7$, i. e. as soon as the primal contains planes; again, in MORIN [2], it is shown that the general quintic primal of S_r is unirational provided that $r \geqq 17$, i. e. as soon as the primal contains spaces S_3. But these results, which are obtained by special methods, are not really particular cases of the theorem about to be proved, for which more stringent conditions are required: in fact we show that the general primal V_{r-1}^n of S_r is unirational for all $r \geqq r_n$, where r_n is defined by the recurrence relation

$$r_n = \binom{n + r_{n-1}}{n}, \quad r_2 = 2\,.$$

For this purpose we require the known result*) that V_{r-1}^n contains spaces S_k, forming an aggregate of dimension D, if, for $n \geqq 3$,

$$D = (r - k)\,(k + 1) - \binom{n + k}{n} \geqq 0\,. \tag{1}$$

It follows then that, if $r \geqq r_n$, V_{r-1}^n must contain spaces $[r_{n-1}]$.

Next, we observe that, writing $r_{n-1} = k$, we readily show that, for $n \geqq 3$, $k \geqq 2$, the inequality $r \geqq \binom{n+k}{k}$ yields

$$r - k - 1 > \binom{n + k - 1}{k}. \tag{2}$$

Now suppose that the general primal V_{s-1}^{n-1} $(s \geqq h)$ is unirational, having a parametric representation of the form

$$\varrho\, x_i = F_i(\xi_1, \xi_2, \ldots, \xi_{s-1}) \quad (i = 0, 1, \ldots, s)\,, \tag{3}$$

*) This is included in a more general theorem, due to PREDONZAN [1], concerning a variety which is a complete intersection of primals.

where F_i are rational functions. Suppose further that the coefficients of the terms in F_i depend rationally on those of the terms in the equation of V_{s-1}^{n-1} and on the parameters which determine a space S_h, where $h = r_{n-2}$, on V_{s-1}^{n-1}: we shall then prove that V_{r-1}^{n} $(r \geqq r_n)$ is likewise unirational, and that the coefficients in the corresponding functions F_i can be made to depend rationally on those of the equation of V_{r-1}^{n} and on the parameters which fix a space S_k on V_{r-1}^{n}.

Let the equation of V_{r-1}^{n} be $f(x_0, x_1, \ldots, x_r) = 0$; then, by a change of coordinates, we can suppose that the S_k in question is represented by $y_i = 0$ $(i = k+1, k+2, \ldots, r)$. The equation of V_{r-1}^{n} then takes the form

$$\sum_{i=k+1}^{r} y_i f_i(y_0, y_1, \ldots, y_k) + F(y_0, y_1, \ldots, y_r) = 0\,, \tag{4}$$

where F is at least quadratic in the variables $x_{k+1}, x_{k+2}, \ldots, x_r$. Evidently the coefficients in this equation are rational functions of the coefficients of $f(x_0, x_1, \ldots, x_r)$ and of the parameters which determine S_k.

Through the general point $(z_0, z_1, \ldots, z_r)$ of V_{r-1}^{n} there passes a unique S_{k+1} containing S_k; this has equations of the form

$$\left.\begin{array}{l} z_{k+1}\, y_{k+2} = z_{k+2}\, y_{k+1} \\ \cdots\cdots\cdots\cdots\cdots \\ z_{k+1}\, y_r \quad = \quad z_r\, y_{k+1} \end{array}\right\} \tag{5}$$

and meets V_{r-1}^{n} in a variety consisting of S_k and a residual V_k^{n-1} which intersects S_k in a V_{k-1}^{n-1} with equation

$$\sum_{k+1}^{r} z_i f_i(y_0, y_1, \ldots, y_k) = 0\,. \tag{6}$$

The latter varies with z_i in a linear system which, by virtue of the generality of the forms f_i, and the inequality (2), coincides with that of all primals V_{k-1}^{n-1} of S_k.

Now choose from the birational system of spaces S_h of S_k a system Σ of dimension k which depends rationally on k non-homogeneous parameters $\xi_1, \xi_2, \ldots, \xi_k$. It follows that a finite number μ of members of Σ lie on the generic V_{k-1}^{n-1} of S_k. In order that a V_{k-1}^{n-1} of the linear system (6) should contain a space S_h of Σ we must impose k linearly independent conditions on $z_{k+1}, z_{k+2}, \ldots, z_r$; that is, the parameters z_i of such a V_{k-1}^{n-1} are expressible as linear functions of $(r-k)-k-1 = r-2k-1$ non-homogeneous parameters $\xi_{k+1}, \ldots, \xi_{r-k-1}$; and the coefficients occurring in these functions are rational functions of $\xi_1, \ldots, \xi_k$. We thus obtain expressions of the form

$$z_i = \varphi_i(\xi_1, \xi_2, \ldots, \xi_{r-k-1}) \qquad (i = k+1, k+2, \ldots, r)\,, \tag{7}$$

where φ_i denotes a rational function.

As the parameters ξ_i vary in (7), we obtain all the primals of (6), each of them μ times. Now, by our inductive hypothesis, V_k^{n-1} possesses the property that the coordinates $(y_0, y_1, \ldots, y_k)$ of its general point can be expressed as rational functions of k non-homogeneous parameters $\xi_{r-k}, \ldots, \xi_{r-1}$, the coefficients in the functions depending rationally on those in the equation of V_k^{n-1} and on the parameters $\xi_1, \ldots, \xi_k$ which fix S_h. We thus obtain the expressions

$$\varrho\, y_i = \psi_i(\xi_1, \xi_2, \ldots, \xi_{r-1}) \qquad (i = 0, 1, \ldots, k+1), \tag{8}$$

where ψ_i denotes a rational function. It now follows at once from (5) that the coordinates of the general point of V_{r-1}^n can be expressed, in the manner prescribed, as rational functions of the parameters.

Following the above procedure, PREDONZAN [2] has proved the more general result that the complete intersection of any number N of general primals of any assigned orders n_i in S_r is unirational provided that r is sufficiently large with respect to N and n_i.

A different line of development is to consider, not linear spaces, but suitable subvarieties of other types on a given variety. Thus MORIN has shown that the general quartic primal in S_6 is unirational, by using the fact that it contains quartic monoidal surfaces (MORIN [8]; PREDONZAN [3]). Other results concerning quartic primals will be found in the next chapter.

4. Varieties containing congruences of rational curves. We consider a non-singular variety V_d which contains a congruence Γ (∞^{d-1} system) of rational curves $\mathfrak{C}$; and we suppose that both the system Γ and the generic curve $\mathfrak{C}$ are irreducible; further, until § 6, we shall assume that Γ has index unity. The congruence may possess base points, simple or multiple, and also singular points, through which pass an infinity of members; again, there may be reducible members, forming a number of aggregates, possibly of various dimensions. To begin with, we can always suppose that the generic curve $\mathfrak{C}$ is non-singular; for if it possessed multiple points, we could apply B. SEGRE's transformation*) a finite number of times and thereby obtain a non-singular variety V'_d, birationally equivalent to V_d, containing a congruence Γ' birationally equivalent to Γ, of rational curves $\mathfrak{C}'$ the generic member of which is irreducible and non-singular. We shall therefore assume that this transformation has, if necessary, already been effected.

We now apply NOETHER's reduction process to the curves $\mathfrak{C}$; for, as ENRIQUES [3] was the first to remark, in the particular case $d = 2$, the reduction is equally valid whether the congruence Γ is birational

*) B. SEGRE [5]; we recall that this transformation is obtained by considering, in the first place, the curve $\mathfrak{C}_1$ corresponding to $\mathfrak{C}$ on the transform W_d of V_d by means of the quadrics of the ambient space and, in the second place, the curve whose points map the tangents to $\mathfrak{C}_1$ on the Grassmannian of the lines of the ambient space.

or not. We thus obtain a birational transform V'_d of V_d which contains a congruence Γ' birationally equivalent to Γ, of conics or twisted cubics, according as the order of the curves $\mathfrak{C}$ is even or odd. In the latter case, we can always construct a unisecant variety*) W_{d-1} to Γ; hence, *if the curves $\mathfrak{C}$ are of odd order, V_d can be transformed birationally into a variety containing a congruence of lines, which is a birational transform of the congruence Γ.*

The situation is essentially different in the case where the curves $\mathfrak{C}$ have even order; for classical examples due to MONTESANO [1] show that even a birational congruence of conics need not possess unisecants; and other examples will be given later.

In the case where Γ is birational, W_{d-1} must obviously be birational; hence, *if Γ is birational and $\mathfrak{C}$ has odd order, V_d is birationally transformable to a space S_d in which Γ is represented by a star of lines.*

Next, if Γ is birational and $\mathfrak{C}$ is a conic, we can transform V_d birationally into a primal W_d of S_{d+1}, of some order n, on which Γ is mapped by the system of conics lying in the planes passing through a fixed line l which is $(n-2)$-ple for W_d; for we have merely to place the system Γ in birational correspondence with the planes through l, and then to project each conic on to its corresponding plane. This operation, in the case $d=2$, was introduced by NOETHER [1], as the first step in the proof of his theorem that any surface which contains a rational pencil of rational curves is itself rational. His method is as follows: assuming that there is a certain number s of distinct nodes of the surface W_2, where $2s < 3n-4$, we shall in general have $3n-2s-4$ line pairs among the pencil of conics whose axis is l; selecting a line from each of $3n-2s-5$ of these pairs, we can construct a curve (necessarily rational) of order $n-2$, $(n-3)$-secant to l, and passing through each of the s nodes of W_2, which meets each of the lines in question. This is the required unisecant. However, it is by no means clear that the construction can be carried out in particular cases, as (for instance) when the locus of double points of the line pairs has components which are in the first neighbourhood of l. This criticism is implicit in COMESSATTI [3], where a purely arithmetic proof of NOETHER's theorem is given (see § 7). The same difficulty re-appears in MORIN [3], where the author seeks necessary and sufficient conditions for the existence of a unisecant to the congruence of conics on W_d by a generalisation of NOETHER's method. Here an essential preliminary is a proof that W_d can be transformed birationally into a primal containing an analogous congruence of conics which satisfies the conditions of generality, enumerated by MORIN, necessary for this line of attack on the problem. Such a proof has yet to be given; assuming that the required transformation can be effected, MORIN has shown that a necessary and

*) If Γ possesses a simple base point, such a unisecant certainly exists.

sufficient condition for the existence of a unisecant to the congruence of conics on W_d (which is supposed not to be a cone) is that the variety which maps the line pairs of the congruence should be reducible, in the sense that each of its components of even multiplicity is to be regarded as virtually non-existent. This condition can readily be translated into one concerning the original variety V_d containing a congruence of curves of even order; on this the prime sections of W_d will be mapped by a linear system of hypersurfaces bisecant to the congruence.

We conclude this section with an interesting proposition, due to FANO [10], concerning threefolds: *if a threefold V_3 contains an irrational (non-birational) congruence, of index unity, of rational curves, it cannot be unirational.* For suppose, if possible, that V_3 can be mapped by an involution I of points in S_3; such an involution may be simple or compound, and the curves $\mathfrak{C}'$ which represent the curves $\mathfrak{C}$ of the congruence Γ in question may or may not belong to I. But, in all cases, since Γ has index unity, the curves $\mathfrak{C}'$ must either form, or be compounded of, a congruence Γ' of index unity. Now Γ' cuts on an arbitrary plane of S_3 an involution of points which, by CASTELNUOVO's theorem, must be birational; from this it follows that Γ' must itself be birational and hence also Γ, contrary to our hypothesis.

5. ENRIQUES's theorem and its applications. Supposing that the birational congruence Γ does not admit a unisecant, it may nevertheless possess a *plurisecant* variety, either birational or unirational: by this we mean an irreducible variety V_{d-1} which is met by the generic $\mathfrak{C}$ in n ($\geqq 2$) distinct and variable points. We then have the result (ENRIQUES [5]): *if Γ admits a birational n-secant variety, then V_d is unirational and representable on an involution I_n.* For, in the first place, we can find a birational representation of V_{d-1}, in consequence of which the coordinates of the general point P of V_{d-1} are expressible as rational functions of $d-1$ essential parameters $(\lambda_1, \lambda_2, \ldots, \lambda_{d-1})$. Again, through P there passes a unique curve $\mathfrak{C}$, rationally determined by P; and, given the coordinates of P, we can obtain a birational representation of $\mathfrak{C}$ by means of which the coordinates of its general point Q are expressed as rational functions of $(\lambda_1, \lambda_2, \ldots, \lambda_{d-1})$ and one other parameter λ_d. Evidently Q is also the general point of V_d. And since the same point Q may be reached in this way from any one of the n points P in which $\mathfrak{C}$ meets V_{d-1}, it follows that there is a $(1, n)$ correspondence between the points of V_d and those of the space in which $(\lambda_1, \ldots, \lambda_d)$ are non-homogeneous coordinates.

This result admits of two obvious extensions which are often useful:

I. In the first place, it is clearly sufficient that V_{d-1} should be *unirational,* in order to prove that V_d is unirational; more precisely, if V_{d-1} is representable on an involution I_m, we infer from the above reasoning that V_d will be representable on an I_{mn}.

II. Secondly, it may happen that the curves $\mathfrak{C}$ are n-secant to a birational or unirational variety V_k, where $k < d-1$. We may then perform a dilatation, i. e. transform V_d birationally by means of primals of any convenient order passing through V_k, and we can then apply the preceding results to this transform. If V_k is unirational and representable on I_m, its transform W_{d-1} by this process will enjoy a like property. To quote a simple example, the planes through a fixed line l of a cubic primal V_d^3 $(d \geqq 2)$ cut the primal residually in a birational congruence of conics which are bisecant to l; hence V_d^3 is representable on an I_2.

We give two interesting applications of these theorems, the first of which concerns the variety V_d which is the complete intersection of m quadrics (non-singular and generically situated) in $[d+m]$. It is obvious (GAUTHIER [1]; ROTH [16]) that V_d is birational provided that the variety contains some space $[m-1]$, since the general $[m]$ through this space must meet V_d in precisely one further point. A necessary and sufficient condition for the existence of such a space on V_d is $d+1 \geqq \binom{m+1}{2}$. However, it may also be proved (ROTH [20]) that *V_d is unirational and representable upon an I_{2m-2} provided that* $d \geqq m^2/2 - 2$, i. e. provided V_d contains some space $[m-2]$. For the sake of simplicity we consider only the case $d=3$, $m=3$, which is due to FANO [11]; the geometrical argument used here, translated into algebraic symbolism, forms the basis of the proof in the general case.

Consider then the V_3^8 which is the complete intersection of three quadrics of [6]; this contains ∞^1 lines. From the generic line l of this aggregate, V_3^8 projects into a quartic primal W^4 of [4], on which the neighbourhood of l is mapped by a cubic scroll R^3. Now through the general point of W^4 can be drawn a unique plane to cut R^3 in a conic; and this plane meets W^4 residually in a second conic which must pass through the point in question, and which is evidently quadrisecant to R^3. We have thus, on W^4, a birational congruence of conics, of index unity, quadrisecant to R^3; from which it follows that V_3^8 is representable on an I_4.

The second illustration is provided by the general quadratic complex $C^2(k, n)$ of subspaces $[k]$ of $[n]$: we shall show that *$C^2(k, n)$ is representable on an involution of order 2^k.* Consider first the case $k=1$, for which the proof is very simple; we may suppose that $n > 3$, since for $n=3$, the variety $C^2(1, n)$ on the Grassmannian $G(1, n)$ is known to be birational. Let l, l' be any two skew lines of $[n]$; since there is a unique [3] containing them, it follows that, on $G(1, n)$, there is a unique quadric $G(1, 3)$ passing through the general point L and also through a fixed point L'. Cutting $G(1, n)$ by a general quadric of the ambient space, which we may assume to pass through L', we obtain, on $C^2(1, n)$, a birational system, of index 1, of Del Pezzo varieties V_3^4 with a simple base point L'. Now, as we have already seen, V_3^4 is unirational in $K(L')$ and representable on an I_2. Hence the same conclusion holds for $C^2(1, n)$.

The general result is now obtained by induction on k. Assuming that the theorem holds for the value $k-1$, we fix a point O of $[n]$ and then determine rationally a $[k-2]$ on the generic $[k]$, e. g. by its intersection with $k-1$ fixed spaces $[n-k]$. Now the $[k]$s passing through this $[k-2]$ and lying in the $[k+1]$ defined by $[k]$ and O form a net, which has a pencil in common with the system of $[k]$s through O; and this system, which is obviously mapped by a Grassmannian $G(k-1, n-1)$, is therefore birational. Cutting the Grassmannian $G(k, n)$ which maps the $[k]$s of $[n]$ by a quadric of the ambient space, we obtain, on $C^2(k, n)$, a birational system of conics, of index 1, bisecant to a fixed $C^2(k-1, n-1)$, which, by the inductive hypothesis, is unirational and representable on an involution of order 2^{k-1}. It follows, then, from the criterion II above that $C^2(k, n)$ is representable on an involution of order 2^k. This result is given in Roth [22].

There is another test for unirationality which has been given incidentally by B. Segre [6], though the idea is to be found in early work of Fano [3]. Suppose that we are given, on V_d, two birational congruences Γ_1, Γ_2, each of index 1, of rational curves $\mathfrak{C}_1, \mathfrak{C}_2$. Fix any curve $\mathfrak{C}_1$; then the ∞^1 curves $\mathfrak{C}_2$ issuing from the points of $\mathfrak{C}_1$ generate a rational surface V_2. Next, consider the curves $\mathfrak{C}_1$ issuing from the points of V_2; it may happen exceptionally that V_2 is generated by ∞^1 of these curves but, if this is not the case, there will be ∞^2 such curves, generating a threefold V_3 which is certainly unirational. Now consider the curves $\mathfrak{C}_2$ issuing from the points of V_3. Proceeding thus, we shall "in general" succeed in constructing a unirational variety V_{d-1} to which the curves of either Γ_1 or Γ_2 are plurisecant, and in that case V_d is unirational.

6. Congruences of index greater than unity. Consider now the case where the congruence Γ carried by V_d has index $\nu > 1$. We first remark that, in the theory of surfaces, the corresponding situation is of no particular interest: thus, if a surface contains a rational ∞^1 system, of index $\nu > 1$, of rational curves, the surface must be unirational, and hence birational; this result, due to Castelnuovo (a), was almost immediately extended by Humbert [1] to the case of any algebraic system. Humbert's original proof was transcendental, but subsequent advances made by Castelnuovo and Enriques in the theory diminished the interest of this result; in point of fact, the virtual grade n and virtual genus p of the system in question satisfy the inequality $n > 2p-2$, so that the carrying surface must be rational or scrollar; but it cannot be scrollar, since the only system of rational curves lying on such a surface is an irrational pencil.

In the theory of manifolds of higher dimension, however, no such simplification is possible: consider, for example, a V_3 which contains an algebraic congruence of lines to which belongs an irrational pencil of quadrics – such a threefold is easily constructed by taking a congruence of index 2. Evidently V_3 must be superficially irregular and hence, as we shall see, it cannot be unirational. Thus the CASTELNUOVO-HUMBERT theorem is not in general extensible to varieties of higher dimension.

We may also remark that FANO's theorem (§ 4) is not in general valid for $d \geqq 3$, $\nu > 1$; for example, the non-singular cubic primal of S_4 contains an irreducible congruence, of index 6, of lines which, as FANO himself has shown, is irregular; nevertheless, the primal is unirational.

Turning to the variety V_d containing the congruence Γ, of index $\nu > 1$, of rational curves $\mathfrak{C}$, we begin by applying SEGRE's transformation (§ 4), thereby obtaining a variety V'_d which, if V_d is non-singular, is likewise non-singular, and which is generated by a congruence Γ', of index 1, of rational curves $\mathfrak{C}'$, birational images of $\mathfrak{C}$. The congruence Γ' is a birational transform of Γ; but V'_d is in $(\nu, 1)$ correspondence with V_d. We may suppose that the generic $\mathfrak{C}'$ is non-singular since, if it possessed multiple points, they could be resolved by applying the SEGRE transformation a finite number of times.

Now whenever we can assert that V'_d is birational, it will follow that V_d is unirational and representable on an involution I_ν. Thus this property will hold provided that Γ' is birational and admits a unisecant. Again, if Γ' is birational and admits a unirational (or, in particular, birational) plurisecant, V_d will be unirational, since it is mapped by an involution on a unirational variety V'_d.

In this and preceding work the question arises whether a birational congruence, of index 1, of rational curves, necessarily admits a unirational plurisecant. This question has not so far been answered: it has, however, been conjectured that the primal V_3^m of S_4 $(m > 4)$ whose only singularity is an ordinary $(m-2)$-ple line l is not unirational; if this were confirmed it would follow that the congruence of conics lying in the planes through l does not admit a rational plurisecant surface.

7. The general problem of unisecants. We have already referred to the proof of NOETHER's theorem presented by NOETHER himself; in this section we shall show that a far more general result concerning unisecants, which includes NOETHER's theorem as a special case, may be established by quite simple algebraic methods.

It will be convenient to begin by considering NOETHER's theorem from this new point of view which, it would seem, was first put forward by BAKER (a), although merely in outline. As remarked in § 4, we can

4*

always reduce the problem to that of constructing a unisecant to the pencil of conics residual to an $(n-2)$-ple line of a surface V_2^n in S_3. The equation of V_2^n, in non-homogeneous coordinates, may be taken to be

$$a_{11}x^2 + a_{12}xy + a_{22}y^2 + a_{23}x + a_{31}y + a_{33} = 0\,,$$

where the coefficients a_{ij} are polynomials in z. Now let u, v be polynomials of order $n-2$, and w a polynomial of order $n-3$, in z; then the equations $x = u/w$, $y = v/w$, represent a rational curve of order $n-2$. The condition that this curve should lie on V_2^n is expressed by the vanishing of a polynomial of order $3n-6$; and there are $3n-5$ disposable constants in u, v, w by choice of which the condition may be satisfied.

A detailed investigation was later given by COMESSATTI [3], who apparently was unaware of BAKER's work. COMESSATTI first remarks that, except in trivial cases where the existence of a unisecant is obvious, V_2^n may be transformed birationally to the surface whose equation is

$$y^2 = Ax^2 + B\,,$$

where A and B are polynomials in z. The unisecant curve will then be obtainable if it can be shown that there exist polynomials u, v, w such that

$$Au^2 + Bw^2 = v^2\,.$$

The existence of these polynomials is then established by counting constants.

The problem of constructing a unisecant on a surface containing an irrational pencil of rational curves had been previously solved by ENRIQUES [3]. Here again, if the curves have odd order, the existence of a unisecant is assured without calculation while, if they have even order, the first step consists in transforming them to conics. However, from this point onwards ENRIQUES's method is long and obscure; in TAGG [1], which is almost contemporaneous with COMESSATTI [3], it is replaced by the algebraic procedure suggested by BAKER's work, though to this the author does not refer. Shortly afterwards, CONFORTO [1] solved the more general problem of finding a sufficient condition for the existence of a unisecant to a congruence, of index unity, of quadric loci of any dimension on a given variety; this includes NOETHER's and ENRIQUES's results as particular cases, and also shows incidentally that, on any threefold, a pencil (rational or irrational) of quadric surfaces always admits unisecants. Later BALDASSARRI [1] extended CONFORTO's result to congruences of hypersurfaces, obtaining not only an arithmetic condition of sufficiency but assigning a condition of "generality" to the congruences which, when satisfied, renders the arithmetic condition necessary as well. As regards the sufficiency, all the above results are included in the following general theorem, due to B. SEGRE [9]:

Let Σ be an irreducible algebraic ∞^d system of varieties V, defined over any field K of constants; and suppose that the generic V lies in a space S_r (fixed or variable), being obtained therein as the complete intersection of a certain number k $(\geqq 1)$ of primals, of given orders $n_1, n_2, \ldots, n_k (\geqq 2)$. Then, if $r \geqq n_1^d + n_2^d + \cdots + n_k^d$, the system Σ admits a unisecant in some extension K' of K. That is to say, there exists a variety W', related to Σ by a rational transformation (defined in K') such that the generic V contains the corresponding point of W'.

We may suppose that the varieties V lie in a fixed space S_r, since we can always reduce to this case by projection. Let Σ be mapped by the primal whose equation, in homogeneous coordinates, is

$$\varphi(u_0, u_1, \ldots, u_{d+1}) = 0\,,$$

where φ is a form of order μ, with coefficients in K; then the generic V is represented by a system of equations, in homogeneous coordinates, of the type

$$f_1(x_0, x_1, \ldots, x_r; u_0, u_1, \ldots, u_{d+1}) = 0, \ldots,$$
$$f_k(x_0, x_1, \ldots, x_r; u_0, u_1, \ldots, u_{d+1}) = 0\,,$$

where f_l $(l = 1, 2, \ldots, k)$ is a form of order n_l in x_i whose coefficients are forms of order ν_l in u_i, with coefficients in K.

Let m denote an integer $(\geqq \mu)$ which we shall choose later, and let h be defined by the expression

$$h = \binom{m+d+1}{d+1} - \binom{m-\mu+d+1}{d+1}.$$

Then we can determine h forms of order m in u_i and with coefficients in K, namely

$$g_1(u_0, u_1, \ldots, u_{d+1})\,, \ldots, \qquad g_h(u_0, u_1, \ldots, u_{d+1})\,,$$

which are linearly independent, mod. φ. Our theorem will be established if we show that it is possible to determine, in some extension K' of K, a set of constants c_{ij}, not all zero, such that the form of order $m\,n_l + \nu_l$ which we obtain by substituting for x_i in $f_l(x_0, x_1, \ldots, x_r; u_0, u_1, \ldots, u_{d+1})$ the expressions

$$x_i = \sum_{j=1}^{h} c_{ij}\, g_j\, (u_0, u_1, \ldots, u_{d+1}) \qquad (i = 0, 1, \ldots, r)$$

is divisible by φ. That is, we must have the identities

$$f_l(x_0, x_1, \ldots, x_r; u_0, u_1, \ldots, u_{d+1})$$
$$= \varphi(u_0, u_1, \ldots, u_{d+1})\, \psi_l(u_0, u_1, \ldots, u_{d+1}) \qquad (l = 1, 2, \ldots, k)$$

where $\psi_l(u_0, u_1, \ldots, u_{d+1})$ is a form of order $mn_l + \nu_l - \mu$ whose coefficients we denote by $a_{l_1}, a_{l_2}, \ldots, a_{l_{h_l}}$, and where

$$h_l = \binom{m n_l + \nu_l - \mu + d + 1}{d + 1}.$$

Now the above identity is equivalent to $k_l = \binom{mn_l + \nu_l + d + 1}{d + 1}$ equations in a_i and c_{ij}; each of these equations has coefficients in K, and expresses the fact that a linear form in a_i is equal to a form of order n_l in c_{ij}. Altogether, then, we obtain $p = k_1 + k_2 + \cdots + k_k$ non-homogeneous equations in a_i and c_{ij}, satisfied by non-zero values of the variables, and the total number q of the latter is given by

$$q = h_1 + h_2 + \cdots + h_l + (r + 1)\, h\,.$$

We shall show that, subject to the inequality stated for r, and for all sufficiently large values of m, we have $q > p$. In that case, interpreting a_i, c_{ij} as non-homogeneous coordinates in S_q, the above p equations represent an algebraic variety containing the origin and such that none of its components can have dimension less than $q - p$. Thus there is at least one component containing the origin, and hence not lying entirely at infinity, having dimension at least $q - p \geqq 1$, and therefore not consisting entirely of the origin.

It follows that we can determine, in some extension K', a point of such a component, not at infinity or at the origin; and this is the desired point.

To complete the proof we remark that, if α, β, γ are any three positive integers such that $\alpha > \gamma$, $\beta > \gamma$, then

$$\binom{\alpha + \beta}{\gamma} = \binom{\alpha}{\gamma} + \binom{\alpha}{\gamma - 1}\binom{\beta}{1} + \binom{\alpha}{\gamma - 2}\binom{\beta}{2} + \cdots + \binom{\beta}{\gamma}.$$

Hence, if α tends to infinity while β and γ remain fixed, we have, with an obvious notation,

$$\binom{\alpha + \beta}{\gamma} - \binom{\alpha}{\gamma} \sim \binom{\alpha}{\gamma - 1}\binom{\beta}{1} \sim \frac{\beta}{(\gamma - 1)!}\,\alpha^{\gamma - 1}.$$

Applying this result repeatedly we obtain for the numbers h, $k_l - h_l$ the approximations

$$h \sim \frac{\mu}{d!}\, m^d\,, \qquad k_l - h_l \sim \frac{\mu}{d!}\, m^d\, n_l^d\,.$$

It now follows that

$$q - p \sim \frac{\mu}{d!}\, m^d (r + 1 - n_1^d - n_2^d - \cdots - n_k^d)\,.$$

Hence, if $r \geqq n_1^d + n_2^d + \cdots + n_k^d$, we have $q > p$, as required.

Supposing, then, that Σ is a congruence of index unity on a variety W, we infer that, whenever r satisfies this inequality, Σ admits a unisecant variety. In particular, when V is a quadric V^2_{r-1}, W is birationally equivalent to a locus simply generated by spaces S_{r-1} (CONFORTO [1]). Again, if V is a SEVERI-BRAUER variety (B. SEGRE, a), and therefore birational in $K(P)$, a like conclusion will hold. Another application, which generalises CONFORTO's result in a different direction, has been noted by BALDASSARRI [1]: when $k = 1$, we have a sufficient condition for the rational transformability of W into an $(n-1)$-ple locus of linear spaces, where n denotes the order of V. Moreover, in the case where Σ is birational, these spaces may be mapped by a star in a linear space of dimension equal to that of W.

In another work, also prior to B. SEGRE [9], BALDASSARRI [3] has discussed the case where W is a primal containing a pencil, rational or irrational, of hypersurfaces endowed with singularities. Let the generic hypersurface V^n_r of the pencil possess an ordinary s-ple variety of dimension $r-1$ which is the complete intersection of V^n_r with another variety of dimension r which contains it simply: then, using methods similar to the above, BALDASSARRI shows that, if $r + 2 > n/s$, W is birationally equivalent to an $(n-1)$-ple primal containing a pencil of spaces S_r, images of V^n_r. If, further, V^n_r is birational in $K(P_i)$, where P_i denotes a finite number of points of V^n_r, the same condition on r suffices to secure that W should be birationally transformable to a cone whose generators S_r are the images of V^n_r.

Yet another generalisation of CONFORTO's result follows by taking $n_1 = n_2 = \cdots = n_k = 2$, and remarking that, if V is non-singular, and of sufficiently high dimension, V projects birationally from the generic S_{k-1} of the system contained in the variety (§ 5). In this case B. SEGRE [9] shows that, if $r \geqq 2^d k + k^2 - 1$, and the generic V of the system Σ is non-singular, then W is birationally equivalent to a locus of linear spaces, images of the varieties V.

8. Varieties containing systems of elliptic or hyperelliptic curves.

An important criterion for the unirationality of a given variety is obtained by extending a method first used by CASTELNUOVO (a) for surfaces which contain a net of elliptic or hyperelliptic curves. Consider, in the first place, a threefold V containing a system Σ of ∞^3 elliptic curves $\mathfrak{C}$, of order n, invading V, and such that the ∞^1 curves $\mathfrak{C}$ which pass through the general point P of V are in general irreducible, forming an irreducible birational system. We suppose for the present that Σ is *free*, i. e. that the passage of a curve $\mathfrak{C}$ through an assigned point does not in general entail its passage through other points; we also suppose that the curves of Σ do not generate a pencil of surfaces.

On the generic curve $\mathfrak{C}$ through P there is a unique point Q such that $(n-1)P + Q$ is a set of the series cut on $\mathfrak{C}$ by primes of the ambient space; we call Q the tangential, or conjugate, of P. It should be noted that Q is rationally determinable when P is given; that the relation between P and Q is asymmetrical, and that the process of taking successive tangentials is in general non-terminating. Also it is clear that the general point of $\mathfrak{C}$ is the tangential of a finite (non-zero) number of points P.

Consider now the locus of Q as $\mathfrak{C}$ moves in the system of curves through P; evidently we obtain an irreducible curve, $V_1(P)$, say, which is birational in $K'(P)$, where K' denotes the extension of K in which Σ is defined. This curve may have a simple or multiple point at P, or not pass through P at all. We then proceed to construct the curve $V_1(P')$ corresponding to the general point P' of $V_1(P)$. There are two possibilities:

1. $V_1(P')$ varies with P'; in this case the locus of the curve is a surface $V_2(P)$ which is unirational, though not necessarily birational, in $K'(P)$.

2. $V_1(P)$ does not vary with P'.

Obviously case 2 cannot occur if $V_1(P)$ passes multiply through P. But it cannot occur even if $V_1(P)$ passes simply through P; for since $V_1(P)$ meets the generic curve $\mathfrak{C}$ through P in just one further point, we should then have the result that, on each $\mathfrak{C}$, the relation between a point and its tangential is symmetrical. Hence, in 2, $V_1(P')$ must be distinct from $V_1(P)$. It is easily shown (Roth [21]) that, in this case, the curves $V_1(P)$ form a congruence of index unity; in any case, since, by hypothesis, the curves $\mathfrak{C}$ invade V, the points Q must likewise do so, and hence there cannot be fewer than ∞^2 curves $V_1(P)$.

Continuing with case 1, we construct the curve $V_1(P'')$ corresponding to the generic point P'' of $V_2(P)$. Again there are two possibilities:

3. The ∞^2 curves so obtained invade V; hence V is unirational in $K'(P)$.

4. As P'' moves over $V_2(P)$, the curves generate a surface $V_2(P'')$; it is readily seen that this cannot happen if $V_1(P)$ passes even simply through P and that, when it does, the surfaces $V_2(P)$ generate an elliptic pencil.

In conclusion, then, the curves $V_1(P)$ either form *a congruence of index unity or an* ∞^3 *system*, and in the latter case they either generate *an elliptic pencil of rational surfaces* or yield *a unirational representation of* V *in* $K'(P)$. Moreover, this last possibility must occur whenever $V_1(P)$ passes at least *simply* through P.

We have so far assumed that Σ is free; if, however, Σ is tied, the same conclusions hold provided that the tangentials of the point P do not fall at one and the same point.

As applications of this method we consider first the threefold V_3^6 which is the complete intersection of a quadric and a cubic primal of S_5 (ENRIQUES [6]); through each point P of V_3^6 there pass two rational ∞^1 systems of plane cubics, lying in planes of the quadric, and belonging to an aggregate of ∞^3 such cubics. Let Σ denote one of the latter; then, with the previous notation, we at once see that $V_1(P)$ has a quadruple point at P, from which it follows that V_3^6 is unirational; a fairly simple calculation shows that V_3^6 is thereby representable on an I_{216}. Instead, however, of moving P' along $V_1(P)$, it is more convenient to move it along one of the ∞^1 lines of V_3^6; this has the effect of lowering the order of the representative involution to 36 (APRILE [1]).

Consider, in the second place, a double space S_3 with a general quartic branch surface; the double lines of S_3 map a system of elliptic curves and, of these, the lines which meet a fixed line form a tied system Σ, to which the previous considerations apply. Hence the double space*) is unirational in $K'(P)$.

Turning now to a variety V_d $(d > 3)$ which contains a system Σ of ∞^d elliptic curves $\mathfrak{C}$ such that those curves of Σ which pass through the general point P form an irreducible birational system, we find that, in order to apply the above method, we must assume that Σ is *unrestricted*, i. e. as P moves over a V_k $(2 \leq k < d-1)$, the corresponding curves $\mathfrak{C}$ generate a V_{k+2}. We may then prove (ROTH [21]) that one of the following possibilities must occur:

1. V_d is unirational in $K'(P)$.

2. V_d contains a congruence, of index 1, of curves V_1 birational in $K'(P)$.

3. V_d contains a congruence, of index 1, of surfaces V_2 unirational in $K'(P)$. . . . .

$(r+1)$. V_d contains a pencil of varieties V_{d-1}, unirational in $K'(P)$.

If $V_1(P)$ passes through P, then case 1 must occur.

We note also that, if case 3 occurs when $d = 3$, then V_3 is generated by an elliptic pencil of surfaces V_2; that, if case 4 occurs when $d = 4$, then V_4 is generated by an elliptic pencil of V_3's; and so on.

Analogous considerations apply to any variety V_d containing a more ample system of elliptic curves endowed with similar properties; with each increase in the dimension of the system comes a reduction in the number of possible cases (ROTH [18]). To take the extreme instance, suppose that through P there passes an irreducible birational system of ∞^d irreducible elliptic curves $\mathfrak{C}$ which invades V_d and which has P for simple base point. Consider the locus of the tangential Q of P in regard to these curves; in general this is a variety W_d, birational in $K'(P)$. Of the ∞^d curves $\mathfrak{C}$ through P, ∞^1 will pass through a second assigned point P' and, of these, a finite number $\nu \geq 1$ will be such that

*) Analogous reasoning shows that the double plane with general quartic branch curve is unirational in $K'(P)$.

the tangential of P falls at P'. We thus obtain a representation of V_d, on an involution I_ν, which is unirational in $K'(P)$. Exceptionally, however, it may happen that the locus of Q is a V_{d-1}; this case will arise if, and only if, the ∞^1 curves $\mathfrak{C}$ passing through P and Q are such that all the tangentials of P fall at Q. It may then be shown (ROTH [21]) that either V_d is unirational in $K'(P)$ or else that it is generated by an elliptic pencil of birational V_{d-1}'s.

Suppose now that the free, unrestricted system Σ on V_d is replaced by an analogous system of *hyperelliptic* curves $\mathfrak{C}$, of genus $p > 1$, but that the other hypotheses remain unchanged. Given the general point P of $\mathfrak{C}$, we can deduce rationally the conjugate Q in regard to the unique g_2^1 contained in $\mathfrak{C}$; in this case the relation between P and Q is symmetrical, a fact which modifies our conclusions. As before, we construct the locus $V_1(P)$ of Q, which is again birational in $K'(P)$, and then the locus $V_1(P')$ corresponding to the points P' of $V_1(P)$, and so on. In this case the hypothesis that $V_1(P)$ passes *multiply* through P will ensure that V_d is unirational in $K'(P)$, but the alternative possibilities are more numerous than before (ROTH [21]).

It is clear that the elliptic and hyperelliptic curves which appear in the above constructions are only the simplest examples of a whole series of varieties to which these considerations can be applied. It is obviously sufficient that the varieties in question should contain series of point-sets analogous to those which we have used on the elliptic and hyperelliptic curves; in that case like conclusions will follow.

9. Threefolds of given curve section. The following results in the theory of surfaces are classical: a) every surface with rational curve sections is either a rational scroll or a Veronese surface (possibly non-normal), b) every surface with elliptic curve sections is either rational (in which case it is a DEL PEZZO surface) or else is an elliptic scroll, c) every surface with hyperelliptic curve sections is either rational (in which case it contains a rational pencil of conics) or ruled. The rational surfaces c) have been classified by CASTELNUOVO (a).

Using these results, ENRIQUES [4] has classified in detail the threefolds with rational or elliptic curve sections and, in broad outline, those with hyperelliptic curve sections. Suppose, in the first place, that V is a threefold with *rational* curve sections; if they are conics, then V is a quadric while, if the surface sections of V are scrolls but not quadrics, V is generated by a pencil (necessarily rational) of planes. Finally, if the surface sections are not ruled, V – or the projection of V on S_4 – is a quartic primal with three double planes meeting in a triple line; on writing the equation of this primal we see that it must be a cone; it is thus the projection of a cone which projects a VERONESE surface from an external point ("VERONESE cone"). Hence in all these cases V is birational.

To obtain the representations of V on S_3 it is sufficient to project from a finite number of points of V. We then find that, in the first case, V is mapped by quadrics through a base conic; in the second, that V is mapped by scrolls of order n, say, with an $(n-1)$-ple base line l, and a number of simple base lines incident to l and skew to each other; in the third case the representative system may be simplified, by an appropriate Cremona transformation, to the system of quadrics touching a fixed plane at a fixed point.

Now suppose that V is a threefold of order n $(n > 3)$, with *elliptic* curve sections. If the surface sections are ruled, V must be generated by an elliptic pencil of planes, since the variety cannot contain ∞^3 lines unless it is a quadric. If, instead, the sections are DEL PEZZO surfaces of the first species, they can be projected birationally on to a plane from a curve of order $n-3$ while, if they are of the second species (in which case $n = 8$), they project from a rational quartic into a quadric. It follows, then, that V can be projected from an appropriate curve into either a space S_3 or a quadric, and is therefore birational.

If V is of the first species, the representation on S_3 will be by cubic surfaces, for the general prime section meets the curve of the projection in three free points. There will be a base curve of order $9-n$, lying on a quadric which is fundamental for the representation. The various possible types are deducible from the classification, by SEMPLE [1], of the threefolds representable on S_3 by systems of cubic surfaces; excluding the cases in which V is a cone, we find that there are the following types:

a) $n = 4$: a base quintic of genus 2.

b) $n = 5, 6, 7, 8, 9$: a base curve consisting of $9-n$ lines through a double base point, the tangent cone at which is fixed.

c) $n = 5$: a rational base quartic.

d) $n = 6$: a base curve consisting of three skew lines.

e) $n = 6$: a double base point, and a base twisted cubic passing simply through it.

f) $n = 6$: a biplanar base point, with fixed tangent plane, and a base cubic lying in this plane and having a node at the base point.

If V is of the second species, and is not a cone, its projection from a rational quartic is a quadric primal. Prime sections of V are mapped by quartic surfaces which are the intersections of this primal with other quadrics. It follows that V can be represented on S_3 by the system of all quadric surfaces.

Finally, we may add that SCORZA [1] has completed the classification of the varieties V_k^n with elliptic curve sections which are neither conical nor generated by a pencil of $[k-1]$s. The main types, from which the rest are obtainable by projection or section, are: the complete intersection of two quadrics of S_{k+2} $(k > 3)$; the Grassmannian V_6^5 of the lines of S_4; the SEGRE V_4^6 which maps the product of two planes; and a V_4^6 which is the intersection of a cone projecting a Veronese surface F^4 from a plane π, with a quadric primal through π and through the cone which projects from π a conic of F^4.

Thirdly, suppose that V has *hyperelliptic* curve sections of genus $p > 1$; if the surface sections are ruled, then V is generated by a pencil, of genus p, of planes. Leaving this case aside, we shall show that V *contains a unique (rational) pencil of quadrics,* which trace the g_2^1 on the general curve section. For consider a pencil of prime sections $|F|$, say, whose base is a curve section $\mathfrak{C}$ of V; through the general point P of $\mathfrak{C}$ there passes a unique conic $\mathfrak{K}$ on each surface F, and these conics all trace the same pair of points on $\mathfrak{C}$, namely P and its conjugate. Hence the locus of the conics $\mathfrak{K}$ is a quadric Q, for any prime of the pencil cuts it in a conic and nothing more. The quadrics Q so constructed form a system of index unity and obviously cannot belong to an ampler system, since F contains only a pencil of conics. Moreover the system is unique, since the pencil on F is unique. We have therefore a unique pencil $|Q|$ on V.

Now, by § 7, $|Q|$ possesses a unisecant curve; hence V is birational, and representable on S_3 in such a way that $|Q|$ corresponds to the pencil of planes through a fixed line l. It follows that prime sections of V can be mapped by surfaces of some order n which have l for $(n-2)$-ple base line and, possibly, other base elements.

The classification of the threefolds V, in the particular case $p = 2$, has been carried out by MORIN [5], who finds that there are 25 distinct families; those types which are not cones have order 10 or less, but there exist also cones of order 11 or 12.

The varieties H_k^n which have hyperelliptic curve sections of genus p, and which are not generated by pencils of $[k-1]$s, have been studied by DU VAL [1], who shows that H_k^n contains a unique pencil $|Q_{k-1}^2|$ of quadrics whose ambient spaces $[k]$ generate a birational variety R_{k+1} of order $n - p - 1$. DU VAL also obtains a series of inequalities for the characters n, p, k, from which he establishes the following theorem: *a necessary and sufficient condition for the existence of a* H_k^n *on which the general* Q_{k-1}^2 *is non-singular is* $(k-1)n \leq 2k(p+1)$.

We conclude this section with a brief account of other classifications by sectional genus. To begin with, it is well known (CASTELNUOVO, a) that a surface of sectional genus 3, if not ruled or a quartic, must be rational or elliptic scrollar; using these results, MORIN [6] has classified*) the threefolds of sectional genus 3, showing that those with scrollar surface sections are all cones. It follows from FANO's theorem (§ 11)

*) A more detailed account of the non-hyperelliptic types is given in JONGMANS [2], evidently written in ignorance of MORIN [6].

that those with rational surface sections are birational – in point of fact, the majority have hyperelliptic curve sections, so that this classification continues the work of MORIN [5].

The non-singular surfaces of sectional genus 4 have been classified by ROTH [7]; those with singularities have been determined by JONGMANS [3], who in the same work has proceeded to classify the threefolds of sectional genus 4.

10. Threefolds containing systems of rational surfaces. We consider now a non-singular threefold V which contains an irreducible system, of given dimension, of rational surfaces, the generic member of which is non-singular. The first group of results, due mainly to ENRIQUES [1], is concerned with systems of dimension three at most; here the basic fact is that V can be transformed birationally, in K, to a threefold containing a like system of surfaces which are one of the four standard types described in § 2.

We begin with the chain of deductions from § 2; these furnish sufficient conditions for the unirationality or birationality of V. Since, however, it cannot be asserted that the majority of the results of § 2 are in any sense best possible, the theorems we obtain may be capable of improvement.

The first of these theorems is as follows: *if V contains a pencil, of genus $p \geqq 0$, of rational surfaces of the first family, it is birationally equivalent to a threefold generated by a pencil, of genus p, of planes.* For we may transform V birationally to a threefold W which contains a pencil, of genus p, of surfaces with rational curve sections; then, on any prime section of W, we have a pencil, of genus p, of rational curves, and this certainly possesses a unisecant curve. Also any surface of the first family is birational in $K(P)$; whence the result. In the case $p = 0$, the pencil of planes may be taken to be in S_3, and V is then birational.

Next, suppose that V contains a *rational pencil of surfaces of the second family*; then V is either birationally equivalent to a threefold W containing a rational pencil of DEL PEZZO surfaces F^n $(3 \leqq n \leqq 9)$ or is representable on a double space S_3, in which the pencil in question is mapped by a pencil of double planes with quartic branch curves; so that the branch surface of the representation is a surface F^{2m} $(m > 2)$ endowed with a $(2m-4)$-ple line and, possibly, other singularities. Evidently a like conclusion holds in the case $n = 4$ (§ 2).

Since, as we have seen, F^n is birational in K when $n = 5, 7, 8$ (first species), it at once follows that V is birational in these cases. This result is true also when F^8 is of the second species; for, referring to the plane representation by curves $\mathfrak{C}^4\,(O_1^2 O_2^2)$, we observe that, by means of the curves corresponding to the adjoint system $\mathfrak{C}^2(O_1,O_2)$, F^8 can be transformed birationally in K to a quadric. Hence the pencil $|F^8|$ on W possesses a unisecant; and we know that F^8 is birational in $K(P)$.

In the case $n = 3$, V is birationally equivalent to a primal V^m of S_4 having an $(m-3)$-ple plane and, possibly, other singularities. Now it follows from BALDASSARRI's work (§ 7) that the pencil of cubic surfaces residual to this plane admits a unisecant. Also, in the case $n = 4$, SEGRE's theorem (§ 7) shows that the pencil admits a unisecant. Thus, when $n = 3, 4$, V is unirational.

In the case $n = 6$, we have a birational representation on a primal V^m containing an $(m-6)$-ple plane, the residual sections of V^m being general projections of surfaces F^6. Now such a projection has a double curve of order 9 which is the complete intersection of the surface with an adjoint cubic surface; hence again, by BALDASSARRI's results [3], the pencil admits a unisecant and, since F^6 is birational in $K(P)$, it follows that V is birational.

Finally, there remains the case $n = 9$; here we have a representation on a primal V^m with an $(m-9)$-ple plane, the residual sections being projections of surfaces F^9; but no further simplification has so far been obtained.

In conclusion, then, *if V contains a rational pencil of surfaces of the second family, it is*

I. *birational* ($n = 5, 6, 7, 8$); *or*

II. *representable on a double S_3 with a branch surface F^{2m} endowed with a $(2m-4)$-ple line* ($n = 2, 3, 4$); *or*

III. *birationally equivalent to a primal V^m of S_4 with an $(m-9)$-ple plane* ($n = 9$), *the residual sections being projected* DEL PEZZO *novenic surfaces.*

The cases in which V contains a rational pencil of surfaces of either the third or fourth family are quickly dealt with; if the pencil is of the third family, the rational pencils of conics contained in its members form a birational congruence, of index 1, on V. Hence V *can be mapped on a double S_3 whose branch surface F^{2m} has a $(2m-2)$-ple point* (and, possibly, other singularities). If, instead, the pencil is of the fourth family, its members can be transformed to double quadric cones with branch curves $\mathfrak{C}^6$ of genus 4, and these in turn can be referred to double planes whose branch curves are sextics with two consecutive triple points. Thus V *can be mapped on a double S_3 whose branch surface F^{2m} has two consecutive $(2m-3)$-ple points joined by a $(2m-6)$-ple line* (and, possibly, other singularities).

Next, suppose that V contains a *net of rational surfaces*, the generic member of which is irreducible. The cases where these belong to the first family or to the second family, type I or II, present no new feature. In the case where they are of the second family, type III, we consider the surfaces of the net which pass through the general point P of V; these form a rational pencil with P as simple base point. And since, as we have seen, a Del Pezzo F^9 is birational in $K(P)$, it follows that, *if* V *contains a net of* Del Pezzo *novenic surfaces, it is birational.*

Suppose, that the surfaces of the net belong to the third family; then, provided that the characteristic system of the net is not compounded of the conics determinable in K, there will be ∞^1 such conics through the general point P of V; these form a rational aggregate, and so generate a rational surface F. Let $|A|$ be a pencil of surfaces of the net; then the above-mentioned conics contained in them form a birational congruence, of index 1, m-secant ($m \geqq 1$) to F. Hence V is unirational and representable on an involution I_m.

Assuming that the surfaces of the net belong to the fourth family, and that the characteristic system of the net is not compounded of the elliptic curves determinable in K, we shall have ∞^1 such curves passing through the general point P of V and forming a rational aggregate. We may thus apply the considerations of § 8 to these curves; *a priori* there are three possibilities: we may obtain a unirational representation for V, or we succeed in constructing either a congruence, of index unity, of non-singular rational curves, or an elliptic pencil of rational surfaces. Now this last alternative is excluded, for it would imply that V is superficially irregular, whereas we know that V contains a net of generic rational surfaces. If the second alternative holds, we have on V a birational congruence, of index unity, of rational curves – for the congruence is mapped by the involution, necessarily rational, which it cuts on a surface of the net – plurisecant to a fixed rational surface, so that V is unirational.

The last results may be summarised (though imperfectly) in the statement: *if* V *contains a net of rational surfaces whose characteristic system is not compounded of rational or elliptic curves, it is unirational.*

In the case where V contains a *web* (∞^3 system) *of rational surfaces*, further improvements can be effected. Provided that the characteristic curves are irreducible they may even be rational or elliptic, for each surface of the web can be referred rationally to a plane involution. Hence, *if* V *contains a web of rational surfaces whose characteristic system is simple, it is unirational.*

It thus remains to consider the case where V contains *a web of surfaces of either the third or fourth family*, having a characteristic system which is compounded of *the rational or elliptic curves determinable in* K, as the case may be. To this end we use a representation of V which is employed in Roth [24]. Suppose that the web $|A|$ belongs to a complete linear system of freedom r, and that the curves $\mathfrak{C}$ of which its characteristic system is compounded form a congruence Γ. Let $|B|$ be a pencil of surfaces of V, not belonging to Γ, and m-secant to $\mathfrak{C}$ ($m \geqq 1$); and refer projectively the surfaces of $|A|$ to the primes passing through a fixed point O of a space S_{r+1}, and those of $|B|$ to a pencil of primes whose base S_{r-1} does not pass through O. We thereby obtain a representation of V on an m-ple cone W of S_{r+1}, with vertex at O and order n, where n ($\geqq 2$) is the number of curves $\mathfrak{C}$ in a characteristic curve of $|A|$. Since Γ is birational and the pencil of curves $\mathfrak{C}$ on A is rational, it follows that W is birational and, moreover, that W has rational curve sections. In the case where $\mathfrak{C}$ is rational we can assert that W is normal ($r = n + 1$); for it will be shown (V, 2) that V is then superficially regular, whence (I, 6) the characteristic system of $|A|$ must be complete. However, we do not require this result in the present section: all we need remark is that W is either generated by a rational pencil of planes or is a VERONESE cone (when $n = 4$) or its projection. Hence either $|A|$ contains two complementary subsystems $|A_1|$, $|A_2|$, each of dimension one at least, and generated by rational pencils $|\mathfrak{C}|$, such that A_1 and A_2 intersect in a single curve $\mathfrak{C}$; or (when $n = 4$) $|A|$ contains a net $|A_1|$ with characteristic curve $\mathfrak{C}$, such that $A \equiv 2A_1$.

We now show that, *if $\mathfrak{C}$ is rational and A belongs to the third family, then V is unirational provided A_1 or A_2 belongs to either the first or second family*. To begin with, A_1 and A_2 are obviously rational, since each surface contains a rational pencil of curves $\mathfrak{C}$. If, then, A_1, for example, is unirational in K, the result is immediate; if instead A_1 is unirational in $K(P)$, we fix any rational curve on a particular surface A_2, and since this is plurisecant to the pencil $|A_1|$, we thus obtain a unirational representation for V. And we know that, if A_1 belongs to the first or second family, one of these alternatives must hold. Since A has hyperelliptic curve sections, A_1 and A_2 could both belong to the third family, in which case nothing could be inferred. If W is a VERONESE cone, we apply the same reasoning to two pencils selected from the net $|A_1|$.

Finally, we show that, *if $\mathfrak{C}$ is elliptic, and A belongs to the fourth family, then V is unirational.* For, since the two components of a surface A now meet in an elliptic curve $\mathfrak{C}$, one must have arithmetic genus zero and the other arithmetic genus -1, because a surface of arithmetic genus $p < -1$ is necessarily scrollar of genus $p > 1$, and hence cannot contain elliptic curves. It follows that, in the present case, W cannot be a VERONESE cone.

Next, we remark that A_1 and A_2 both have geometric genus zero; for if A_1, say, possessed an effective canonical system, so also would A, by virtue of the relation $|A'| = |A_1' + A_2|$. We may thus suppose that A_1 has genera $p_g = p_a = 0$, and that A_2 has genera $p_g = 0$, $p_a = -1$.

Again, A_1 must have bigenus zero; for, from the formulae (ENRIQUES, a) which give the plurigenera of a surface generated by a rational pencil of elliptic curves, we know that otherwise the pencil $|\mathfrak{C}|$ on A_1 would contain multiple elliptic curves, whereas the congruence Γ, consisting of pencils (determinable in K) on rational surfaces of the fourth family, does not possess such curves. Hence A_1 is rational.

For a similar reason A_2 must be elliptic scrollar; for we know that the rational pencil of elliptic curves on an elliptic surface with invariants $p_g = 0$, $p_a = -1$, necessarily contains multiple members (ENRIQUES, a).

It now follows that A_1 must belong to the first or the second family. For, transforming the system $|A|$ into the system $|F^9|$ referred to in § 2, we see that, when F breaks up into components F_1, F_2, where F_1 is rational and F_2 elliptic scrollar, F_1 and F_2 have in common a plane cubic, so that the respective sectional genera π_1, π_2, of F_1, F_2 satisfy the relation $\pi_1 + \pi_2 = 2$; and since $\pi_2 \geqq 1$, we have $\pi_1 \leqq 1$. If, then, F_1 is unirational in K, V is unirational; if instead F_1 is unirational in $K(P)$, we fix one of the ∞^1 rational curves on a particular surface F_2; whence the required result, as above (ROTH [24]).

11. FANO's theorem and allied results. We have seen in § 10 that, if a threefold V contains a web of rational surfaces, with irreducible characteristic system, it is unirational. In the case where we are given, on V, a completely irreducible system of rational surfaces, of freedom $r \geqq 4$, more than this can be proved. Such a system can be mapped on the prime sections of a threefold W, normal in some space S_s ($s \geqq r$); for this we shall establish the following theorem, due to FANO [8]:

Any threefold whose prime sections are rational is either birational or birationally equivalent to a non-singular cubic primal of S_4.

To begin with, we may always suppose that the threefold W in question is free from singularities; then, since the system $|F|$ of its prime sections is generic in the sense of I, 6, it follows that W is superficially regular, and hence that the characteristic system of $|F|$ is complete: in other words: the surfaces F are normal. Again, it is obvious that the geometric genus and plurigenera of W are all zero; and, since $q_2 = P_g = 0$, it follows from I, 10 that the arithmetic genus P_a satisfies the inequality $P_a \leqq 0$. Actually, of course, we already know that W is unirational; but we cannot conclude from this that $P_a = 0$, i. e. that W is completely regular (see V, 10).

In FANO's proof, which we proceed to describe, the latter fact is fundamental. FANO's method is to consider the system $|(2F)'|$ of surfaces; if this is simple, and sufficiently ample, it can be mapped on the prime sections of another threefold, to which the previous considerations can then be applied; and so on. In this way W may be birationally transformed to one of a series of standard types which, with the possible exception of the non-singular cubic primal, are shown to be birational.

Besides being exceedingly lengthy, this procedure involves the unproved assumption that $P_a = 0$, which FANO attempts to justify as follows*). Leaving aside the cases where W has rational, elliptic or

*) A transcendental proof that any unirational variety V_d has arithmetic genus zero follows from the formula (III, 2) $P_a = g_d - g_{d-1} + \cdots + (-1)^{d-1} g_1$.

hyperelliptic curve sections (and where, by § 9, the result is immediate) we may map W on one of the double spaces obtained in § 2. If the corresponding branch surfaces have no other singularities than those prescribed in the "general" case, a simple calculation shows that $P_a = 0$. But we cannot exclude the possibility that there may arise additional singularities infinitely near to the prescribed ones; and FANO remarks that it is in the nature of the case that these can have no influence on the value of P_a. Actually he makes no more than a brief statement to this effect, but it is difficult to imagine how even a detailed argument could lead to the desired conclusion.

In MORIN [4] a greatly simplified demonstration of FANO's theorem is given, which is based on the classic procedure of successive adjunction, due (in the case of surfaces) to CASTELNUOVO and ENRIQUES. The underlying idea, which derives from I, 10, is as follows. If $|A|$ is any given system of surfaces on a threefold, we know that the system $|A'|$ cuts A in curves of the virtual impure canonical system; hence the system $|A + A'|$ cuts A in curves belonging to the system adjoint to that of the characteristic curves on A; and, generally, the system $|A + mA'|$, where m is any positive integer, cuts A in curves which are m-adjoint to the characteristic system. In the case where the threefold is completely regular, we have seen that $|A'|$ cuts the complete canonical system (if effective) on A; and also (III, 2) that each of the systems $|A + mA'|$ $(m = 1, 2, \ldots)$, if effective, cuts a complete system on A.

Now if A is rational, the system $|A + mA'|$ must be virtual for all sufficiently large values of m; and for the last value n, say, of m for which the system is effective, it will cut either rational or elliptic curves on A. Hence, identifying $|A|$ with $|F|$, and assuming that W is completely regular, we can prove FANO's theorem, obtaining a projective classification of W at the same time.

This is MORIN's method: in order to circumvent the difficulty concerning the unproved hypothesis, we shall show how to prove the theorem by using ENRIQUES's standard forms (§ 2), after which, with the knowledge that $P_a = 0$, we can proceed with the projective classification.

We begin, then, by transforming W birationally so that $|F|$ becomes a system of one of the types listed in § 2; since the generic F is non-singular, and the process is completed in a finite number of stages, we shall thereby obtain a threefold whose generic prime section is non-singular, which means that the threefold can possess at most a finite number of multiple points. If this threefold – which we may continue to call W – has rational, elliptic or hyperelliptic curve sections, the theorem is proved; thus there remain to be considered merely the cases in which the generic prime section of W is

I. the surface F^8 mapped by curves $\mathfrak{C}^6(O_1^2 O_2^2 \ldots O_7^2)$; or

II. the surface F^9 mapped by curves $\mathfrak{C}^9(O_1^3 O_2^3 \ldots O_8^3)$.

Now, as first remarked by CASTELNUOVO (a), the surface F^8, which is normal in S_6, is obtainable as the complete intersection of a VERONESE point-cone and a quadric (not passing through the vertex). From this it readily follows (FANO [8]) that W is the complete intersection of a VERONESE line-cone and a quadric not passing through the vertex. As FANO has remarked, W projects into a VERONESE point-cone from either of the points (which may possibly coincide) where the quadric meets the vertex. Hence W is birational.

Again, it may be shown that F^9, which is normal in S_6, is obtainable as the intersection of the cone K which projects, from an external point, a rational normal quartic scroll endowed with a linear directrix, and a cubic primal which passes through three generating planes of K. Hence, (FANO [8]) W is the intersection of a line-cone which projects, from an external line, the above-mentioned scroll, and a cubic primal of S_7 which passes through three S_3-generators of this cone; whence, by fairly simple geometrical considerations*), it may be shown that W is birational.

Thus in all cases the original threefold V, say, is birational or birationally equivalent to a cubic primal of S_4, and so completely regular. We can therefore assert that, if F and $\mathfrak{C}$ denote respectively the generic surface section and curve section, the system $|F + mF'|$, when effective, cuts on F, residually to possible fixed components, the complete system $|\mathfrak{C}^{(m)}|$ of curves m-adjoint to the system $|\mathfrak{C}|$ on F. And since F is rational, for a certain value n of m, the system $|\mathfrak{C}^{(n)}|$, of freedom $s \geqq 1$, either consists of irreducible rational or elliptic curves or else is compounded of a rational pencil of rational curves. That these curves cannot be elliptic is seen as follows. If $|\mathfrak{C}^{(n)}|$ is a pencil of elliptic curves, F can always be mapped on a plane so as to represent $|\mathfrak{C}^{(n)}|$ by a pencil of plane cubics (hence, incidentally, the system $|\mathfrak{C}^{(n-1)}|$ will be mapped by the ∞^3 system of sextics having 8 double base points and with characteristic series compounded of an involution I_2). If, then, $\mathfrak{C}^{(n)}$ were compounded of a certain number of elliptic curves of a rational pencil, the corresponding set of plane cubics would behave like a reducible member of a HALPHEN pencil; whereas such a pencil, being superabundant, cannot be an adjoint system.

MORIN's analysis, which we now sketch, is based on these considerations. Supposing that the curves $\mathfrak{C}$ have genus 2 at least, two main possibilities present themselves.

I. $\mathfrak{C}^{(n)}$ is either rational (irreducible) or is compounded of a number of rational curves. In either case V contains a rational pencil of surfaces having rational curve sections; in particular, if these surfaces are quadrics, V has hyperelliptic curve sections.

*) FANO proves that this threefold is the projection, from a curve of order 9 and genus 4, of the V_3^{36} considered below.

II. If $\mathfrak{C}^{(n)}$ is elliptic (necessarily irreducible), the surface $F + nF'$ has elliptic curve sections. In the case where $s \geqq 2$, the characteristic curve of $|F + nF'|$ has order s (equal to the order of the characteristic series of $|\mathfrak{C}^{(n)}|$ on F). If $s > 2$, because the complete system $|\mathfrak{C}^{(n)}|$ on F is simple, it follows that, in the case where $|F + nF'|$ is compounded of a congruence of curves, these must be lines, i. e. V is ruled. Supposing this is not the case, we observe that the system $|2(F + nF')'| = |F + (2n + 1)F'|$ is virtual, which means that the characteristic curves of $|F + nF'|$ are rational.

We next show that $s \leqq 4$. For suppose, if possible, that $s > 5$; then the series, of order s, cut by $|\mathfrak{C}^{(n)}|$ on a curve $\mathfrak{C}$ of F would have order greater than 9, in contradiction with the result that a Del Pezzo surface has order 9 at most. Again, if $s = 5$, $n = 1$, $|\mathfrak{C}^{(n)}|$ cuts the canonical series, of order 10, on $\mathfrak{C}$, which is impossible for the same reason. If $s = 5$, $n > 1$, we may map F on a plane in such a way, that $|\mathfrak{C}^{(n)}|$ corresponds to the cubics through four base points P_i. The system $|\mathfrak{C}|$ on F will be mapped by that of curves of a certain order $N \geqq 9$ (since $n > 1$) having multiplicities r_i at P_i, where $\Sigma r_i < 2N$, as we see by considering the conics through P_i. Then the series cut by $|\mathfrak{C}^{(n)}|$ on $\mathfrak{C}$ has order $3N - \Sigma r_i > N \geqq 9$, which again leads to a contradiction.

The cases $s = 3, 4$ are quickly disposed of, leading respectively to threefolds which are mapped by systems of cubic surfaces in S_3 and systems of quadric sections of a quadric primal. The remaining cases $s = 2, 1$ require more detailed consideration.

If $s = 2$, the characteristic curve of $|F + nF'|$ has 2 intersections with F, and is thus either a pair of lines (in which case V is ruled) or a conic; in the latter case we have on V a congruence, of index unity, of conics, $\mathfrak{K}$, say.

Suppose, in the first place, that $n > 1$; then the system $|F + (n - 1)F'|$ cannot be compounded of the conics $\mathfrak{K}$, since the system $|\mathfrak{C}^{(n-1)}|$ is simple. We next observe that the surfaces of this system are rational; for since the dimension of the system is at least 3, and V is superficially regular, $F + (n - 1)F'$ must be regular – and obviously has geometric genus zero, since its adjoints are virtual. Also its bigenus is zero; for if it possessed an effective bi-adjoint $(2nF')$, this would be $2n$-adjoint to F.

Consider, then, the system cut by $|F + nF'|$ on the rational surface $F + (n - 1)F'$; this is the complete system of curves adjoint to the curve sections of $F + (n - 1)F'$. But since the system formed by the surfaces $\{[F + (n - 1)F'] + [F + nF']\}' = F + 2nF'$ is virtual, the variable curve of intersection of $|F + nF'|$ with $F + (n - 1)F'$ is rational. Hence $|F + nF'|$ determines a homaloidal net on $F + (n - 1)F'$ and hence, incidentally, the surfaces $F + (n - 1)F'$ are unisecant to the conics $\mathfrak{K}$.

Further, since the surface $\{2(F + (n-1)F')\}'$ is virtual, the system $|F + (n-1)F'|$ has rational characteristic curves; hence the image of this system is a threefold on which the conics $\mathfrak{K}$ are mapped by lines.

Suppose, in the second place, that $n = 1$; then F is the surface F^8 already described.

Finally, we have to examine the case $s = 1$; here, as we have remarked, the system $|\mathfrak{C}^{(n-1)}|$ consists of sextics and its characteristic series is compounded of an involution I_2. Then the system $|F + (n-1)F'|$, which cuts $|\mathfrak{C}^{(n-1)}|$ on F, is compounded of a congruence, of index unity, of conics $\mathfrak{K}$, or of lines (in which case V is again ruled). Supposing, in the former case, that $n > 3$, we observe that the system $|F + (n-1)F'|$ cuts on the rational surface $F + (n-2)F'$ a complete system of rational curves, by reason of the fact that the surface $\{[F + (n-2)F'] + [F + (n-1)F']\}'$ is virtual. Since this system of curves is simple, the surface $F + (n-2)F'$ is unisecant to the conics $\mathfrak{K}$.

If $n = 3$, we see that the system $|F + 2F'|$ cuts a system of elliptic curves on a surface $F + F'$. Since the system $|F + 2F'|$ has freedom 3 at least, $F + F'$ is unisecant to the conics $\mathfrak{K}$.

If $n < 3$, the system $|F + (n-2)F'|$ has freedom 3 at least, since the system defined by the surface $\{2(F + (n-2)F')\}'$ has rational or elliptic characteristic curves. In the case $n = 3$, we see that $|F + F'|$ has a characteristic curve of genus 2; as Fano has remarked, this curve may be regarded as adjoint to the system cut on $F + F'$ by the system $|F - F'|$; the system of such curves must therefore be simple, since its last adjoints are rational curves. Hence in all cases the system $|F + (n-2)F'|$ is simple; its image is a threefold containing a congruence of lines which map the conics $\mathfrak{K}$. The one remaining case $s = 1$, $n = 2$, gives the novenic threefold described above.

In conclusion, Morin gives the following list of birational threefolds, from which others may be obtained by projection:

1. Threefolds which are generated by rational pencils of planes.
2. Threefolds containing rational pencils of Veronese surfaces.
3. Threefolds generated by rational pencils of quadrics.
4. The V_3^{27} of S_{19} mapped on S_3 by cubic surfaces.
5. The V_3^{16} of S_{13} mapped by quadric sections of a quadric primal of S_4.
6. The V_3^{32} of S_{21} mapped by quadric sections of a Veronese point-cone.
7. The V_3^8 of S_7, intersection of a Veronese line-cone with a quadric of the ambient space which does not pass through the vertex of the cone.
8. The V_3^{36} of S_{22} mapped on the point-cone of S_6 which projects a rational normal quartic scroll with linear directrix, by sections with cubic primals of S_6 which contain three generating planes of the cone.
9. The V_3^9 of S_7, intersection of a line-cone which projects the above scroll, and a cubic primal of S_7 which contains three S_3-generators of the cone.

MORIN gives also the representations on S_3 of each of the above types, and the irrationalities which are met with in effecting such representations. He then proceeds to classify, by the same methods, the varieties of higher dimension whose surface sections are rational; here again a certain hypothesis of regularity is required in the argument.

MORIN's method of successive adjunction forms the basis of the classification, by BALDASSARRI [2], of the threefolds which contain *webs of rational surfaces whose characteristic series are compounded* (of an involution of some order). His final result is that any such web is a subsystem of the prime sections of a threefold which is either birational or birationally equivalent to a cubic primal, the latter possibility occurring only if the given web has elliptic characteristic curve and grade 2 or 3. Here also the hypothesis that a unirational threefold is completely regular is invoked.

BALDASSARRI begins by determining the surfaces of S_3 which possess ∞^2 rational or elliptic plane sections; he shows that the surfaces of the first kind either have elliptic curve sections or possess a proper multiple point, in which case they are normal in S_3 only if they are monoids; otherwise they are projections of surfaces with a proper multiple point from which they project into a VERONESE surface or a rational scroll. Those of the second kind either have curve sections of genus 2 or possess a proper multiple point, in which case they are normal in S_3 only if they have order n and an $(n-2)$-ple point; otherwise they are projections of surfaces with a proper multiple point from which they project into surfaces with elliptic curve sections. All these surfaces are of course rational, and their plane representations are readily found.

Consider now a threefold containing a web of the above type; we may suppose this to be transformed birationally into a primal V of S_4 on which the web is mapped by the system of prime sections passing through a fixed point O which must be a proper multiple point of some order r, say, otherwise any prime section of V would be rational, and we could apply FANO's theorem. We may assume that V is not a cone, since the characteristic system of the web would then be compounded of a congruence. There are two cases to consider, according as the plane sections of V through O are rational or elliptic; and, applying the previous results for the surfaces with ∞^2 curve sections of either type, we find that, in the first case, V is either a monoid or is normal in higher space, from which it may be projected into a threefold with rational curve sections; and that, in the second case, if the plane sections through O have order exceeding $r+3$, V is again birational. When, however, these curves have order $r+2$ or $r+3$, V may be birational or birationally equivalent to a non-singular cubic primal of S_4.

Finally, using the above method of successive adjunction, we may obtain, in the cases where V is birational, the corresponding representations on S_3.

12. Threefolds containing systems of surfaces with $p^{(1)} \leqq 1$. In the theory of surfaces, many problems of rationality are closely related to the properties of surfaces which contain a pencil or net of rational, elliptic or hyperelliptic curves. These results suggest that we may develop the corresponding theory of threefolds in one or other of two directions:

I. the consideration of a threefold which contains an algebraic system, of appropriate dimension, of rational, elliptic or hyperelliptic curves;

II. the consideration of a threefold containing a linear system, of appropriate freedom, of surfaces which are in some sense analogous to such curves.

We have already dealt with I. in some detail: as regards II., we have discussed the case of rational surfaces, which are analogous to the rational curves in that they possess no effective canonical or pluricanonical systems. A second analogy with rational curves is provided by the scrollar surfaces; these may be partially characterised*) by the fact that their absolute linear genus $p^{(1)}$ satisfies the inequality $p^{(1)} \leqq 1$. Again, an obvious analogy with the elliptic curves is furnished by the surfaces, regular or irregular, which possess a pure canonical curve, effective or virtual, of order zero: these have linear genus $p^{(1)} = 1$. In the present section we shall consider threefolds which contain surfaces of one or other of these types; an interesting class of threefold belonging to this category is, however, reserved for the next chapter.

Let V be a non-singular threefold which contains a net $|R|$ of scrollar surfaces of genus $p > 0$ such that the generic R is irreducible non-singular. The characteristic curve of $|R|$ may be simple or compound; in the latter case we shall assume that is not compounded of the generators $\mathfrak{C}$ of R, i. e. the rational curves of which R contains a pencil, of genus p. This being so, it follows that through the general point P of V there pass ∞^1 curves $\mathfrak{C}$, forming an irreducible rational system, since they are in birational correspondence with the surfaces of $|R|$ which have a simple base point at P. These curves generate an irreducible surface A which is obviously birational in $K(P)$. Proceeding as in § 8, we now construct the surfaces A corresponding to the points P of a given curve $\mathfrak{C}_0$, say; there are then two possibilities:

I. A varies with P; in this case A describes a rational pencil.

II. A does not vary with P; then the ∞^2 curves $\mathfrak{C}$ passing through the points P of $\mathfrak{C}_0$ lie on one and the same surface A.

*) Any surface for which $p^{(1)} < 1$ is scrollar; a surface for which $p^{(1)} = 1$ is elliptic scrollar if, and only if, $p_{12} = 0$.

In case I, the ∞^2 curves $\mathfrak{C}$, each of which meets $\mathfrak{C}_0$ in a point P which is simple for $\mathfrak{C}$, invade V. Since each curve $\mathfrak{C}$ is birational in $K(P)$, it follows that V is unirational: in particular, if the system of curves $\mathfrak{C}$ has index 1, V is birational.

In case II, which can arise only if P is simple for A, by moving $\mathfrak{C}_0$ over a surface R, we obtain a system $\{A\}$ which evidently has index 1, and in fact is a pencil of genus p. Every surface A contains a net $|\mathfrak{C}|$ which is cut on it by $|R|$.

Now if the characteristic system of $|R|$ is simple, the surface R is generic in the sense of CASTELNUOVO-ENRIQUES, which means that V must have superficial irregularity p, and so cannot be unirational. In case I, R cannot be generic, so that the characteristic system of $|R|$ is necessarily compound. We thus have the following result (ROTH [27]):

If the characteristic system of the net $|R|$ is simple, V is generated by a pencil, of genus p, of surfaces which are birational in $K(P)$; if the system is compound, V is either generated by such a pencil, or is unirational.

Next, turning to the non-scrollar surfaces for which*) $p^{(1)}=1$, $p_{12}>0$, we observe that all these contain at least one pencil, rational or irrational, of elliptic curves, except for the following types:

a) regular surfaces of genera and plurigenera unity;

b) PICARD surfaces (for which $p_g=1$, $p_a=-1$).

Among these there are, however, particular cases which contain pencils of elliptic curves; we know that, if a surface b) contains an elliptic curve, it must possess two pencils, both elliptic, of such curves. The threefolds whose prime sections are of type a) will be considered in detail in the next chapter; here we shall examine the threefolds which contain a *net* of surfaces of linear genus unity such that *each surface possesses a uniquely definable pencil, rational or irrational, of elliptic curves.*

An important subclass of such surfaces is constituted by the *elliptic surfaces* (ENRIQUES, a); on these the pencil in question is that of the trajectories of the group of automorphisms of the surface (VI, 2). Every irregular non-scrollar surface of geometric genus zero is elliptic – in this case the pencil of trajectories is rational.

Suppose that the threefold V contains a net $|F|$ of such surfaces, and that the characteristic system of the net is not compounded of the elliptic curves $\mathfrak{C}$ in question. In that case the curves form a system to which we may apply the considerations of § 8; with the notation of that section we have, in the first place, the two possibilities:

1. The curve $V_1(P')$ varies with P': then the locus of $V_1(P')$ is a surface $V_2(P)$, unirational in $K'(P)$.

2. $V_1(P')$ does not vary with P': in this case the curves $V_1(P)$ form a congruence, of index 1, unisecant to the surfaces of $|F|$, i. e. V is

*) For a discussion of these surfaces see NOLLET [2].

birationally equivalent to a cone which projects F from an external point and hence, by § 4, is not unirational.

Continuing with the discussion of case 1, we have the two possibilities:

3. The ∞^2 curves $V_1(P'')$ corresponding to the points P'' of $V_2(P)$ invade V: hence V is unirational in $K'(P)$.

4. The curves $V_1(P'')$ generate a surface $V(P'')$: in this case we obtain an elliptic pencil of such surfaces, unisecant to the curves $\mathfrak{C}$, and so V is birationally equivalent to the product of a rational surface and a curve $\mathfrak{C}$.

These results are valid whether F is regular or irregular; when, however, we know something further concerning F, more precise conclusions may be drawn. Thus, if V is unirational, it must be superficially regular; if it is transformable to a cone projecting F, its superficial irregularity is equal to that of F (I, 11), while if it is birationally equivalent to the product of a rational surface and a curve, it has superficial irregularity unity.

Now, in the case where F is regular and generic, V is of course superficially regular, so that the last possibility is excluded. Hence, *if F is regular, and if $|F|$ has a simple characteristic curve, V is either unirational in $K'(P)$ or is birationally equivalent to a cone projecting F.*

If, instead, F is generic and irregular, the first possibility is excluded, and the third also, except in the case where F has irregularity 1, i. e. (ENRIQUES, a) when F is either an elliptic surface with invariants $p_g = 0$, $p_a = -1$, or a surface containing an elliptic pencil of elliptic curves ($p_g > 0$, $p_a \geqq 0$). Hence:

If F is irregular, and if $|F|$ has a simple characteristic curve, V is birationally equivalent to a cone projecting F, except possibly when F has irregularity 1, in which case V may be birationally equivalent to the product of a rational surface and an elliptic curve.

The above results, which are obtained in ROTH [27], are often capable of improvement in cases where we are given a more ample system of such surfaces on V: in particular, where V has prime sections of an assigned character. To begin with, we have the theorem, due to PREDONZAN [4]: *If the prime sections of V are scrollar, then V is planar (i. e. birationally equivalent to a plane-ruled threefold)* *). Assuming that the prime sections have genus p (> 0), we know that V must have superficial irregularity p. And since V contains a net of scrollar surfaces with simple characteristic system, it follows from what has just been proved that V is generated by a pencil, of genus p, of rational surfaces; the essentially new feature in the situation is that these surfaces can be birationally transformed to planes.

*) We say that V is plane-ruled if it is generated by a pencil, rational or irrational, of planes.

Let r and ϱ $(1 \leqq \varrho \leqq r-1)$ denote respectively the dimensions of the ambient spaces of V and the generic rational curve $\mathfrak{C}$; then the totality of such curves has dimension $d = r+1-(r-\varrho-1) = \varrho + 2$. Let F denote the surface generated by the curves $\mathfrak{C}$ passing through the general point of V; such surfaces form an irrational pencil. Suppose, in the first place, that $\varrho = r-1$, so that $d = r+1$; then F has rational curve sections, and is therefore either a rational scroll or a VERONESE surface. Hence, by § 10, we can transform the pencil $\{F\}$ birationally into a pencil of planes. Suppose, in the second place, that $3 \leqq \varrho < r-1$; then, projecting V on to a $S_{\varrho+1}$, we may apply the preceding result. It may further be shown that the case $\varrho = 2$ cannot arise; in the case $\varrho = 1$, there is nothing to prove.

An important type of surface of linear genus unity is the ENRIQUES surface $(p_g = p_a = 0,\ p_2 = 1)$, which is birationally equivalent to the sextic surface of S_3 which passes doubly through the edges of a tetrahedron. This surface in general contains three rational pencils of elliptic curves, so that the previous considerations are not applicable to it: when, however, we are given a threefold whose prime sections are ENRIQUES surfaces, a remarkable simplification ensues. Such threefolds have been studied by FANO [13]; the threefold of lowest order is the primal – which we shall call the ENRIQUES primal – of S_4 whose generic prime section is an ENRIQUES sextic surface. In the next chapter we shall show that this primal is unirational but not birational. The threefolds of higher order are few in number, and FANO shows that they are all birational; since both the demonstration and the classification are intimately connected with the FANO threefolds (V, 4), we shall leave the discussion to a later stage.

13. Linear systems of surfaces of maximum dimension. In a fundamental paper on linear systems of plane curves, CASTELNUOVO (a) proved the theorem:

Any linear system of plane curves of virtual genus $p > 1$, whose freedom exceeds $3p + 5$ is compounded of a pencil of rational curves; and any linear system whose freedom exceeds $2p + 7$ is either of the preceding type or else consists of hyperelliptic curves, except for the systems of quartics ($p = 3$) and quintics ($p = 6$) and their birational transforms.

Later ENRIQUES [8] obtained similar results for linear systems of curves on a surface; in particular, he showed that, if the freedom of a linear system, with $p > 1$, exceeds $3p + 5$, the surface must be ruled. With the subsequent development of the theory, this result lost its significance, becoming absorbed in a more expressive theorem.

In a comment on his own work, CASTELNUOVO (a, p. 186) has proposed the problem of determining the linear systems of surfaces of S_3 which have maximum freedom, at the same time stating some provisional results from unpublished work. The problem has been solved in

PREDONZAN [5], completing previous results of MORIN. PREDONZAN approaches the question by classifying the threefolds V_3^n with regular surface sections, with curve sections of genus p ($\geqq 2$) such that $n \geqq 3p - 3$, and normal in space S_r. Since $n > 2p - 2$, the surface sections of V_3^n are rational, whence V_3^n is birational. Assuming that V_3^n is not a cone, PREDONZAN finds five main types; for the first type, which has hyperelliptic curve sections, $r = 3p + 4$, while, for the remaining types, with non-hyperelliptic curve sections, $r < 3p + 4$. The linear systems of surfaces which map these threefolds on S_3 are also given; and since such a linear system, for which $r = 3p + 4$, maps a threefold of the kind under consideration, the reasoning can be inverted and CASTELNUOVO's problem thereby solved.

In PREDONZAN [4] a result analogous to the above theorem of ENRIQUES is obtained. Let $|F|$ be a linear system, of freedom r, on a superficially irregular threefold V; and suppose that the free characteristic curve of the system is irreducible and of genus $p \geqq 3$, where $r \geqq 3p + 6$. Then we can map $|F|$ on the prime sections of a threefold V', which will be a simple or multiple model of V, according as the characteristic series of $|F|$ is simple or compound.

Now the surface F must be scrollar of genus p. For ENRIQUES [8] has shown that, on any non-scrollar (or non-rational) surface, a linear system of curves of genus p is simple provided its freedom exceeds $2p + 6$. If, then, F were not scrollar, the characteristic system cut by $|F|$ on F would be simple, and hence the system $|F|$ must also be simple. Thus the prime section F' of V', which has sectional genus p and is a birational transform of F, would be non-scrollar; and hence, by another proposition of ENRIQUES [8], we should have $r < 3p + 6$, in contradiction to the hypothesis.

Suppose, first, that $|F|$ is simple; then F' is a birational transform of F, and hence scrollar of genus p; thus, by what has been proved in § 12, V' is planar.

Next, assuming that $|F|$ is compounded of an involution I_m ($m \geqq 2$), and denoting by p' the sectional genus of V', we have, by ZEUTHEN's formula, $m(p' - 1) \leqq p - 1$. And since $m \geqq 2$, it follows that $p' \leqq (p + 1)/2$. Hence, from the inequality $r \geqq 3p + 6$ ($p \geqq 3$), we deduce that $r \geqq 3p' + 8$, so that, by ENRIQUES's results, F' must be ruled; therefore V' is plane-ruled. To a plane of V' there corresponds on V a surface consisting of $m/s \geqq 1$ irreducible components G, where the number s is a divisor of m; and the aggregate of all surfaces G obviously forms a pencil.

If $s = 1$, each of the m surfaces G is birationally equivalent to the corresponding plane of V', and so is rational. To a curve section of V' there corresponds on V a unisecant to the pencil $\{G\}$.

In the case where $s > 1$, it follows from ENRIQUES [8] that $r \leqq 2p + 4$, which contradicts our hypothesis. Hence, in conclusion:

Any superficially irregular threefold which contains a linear system $|F|$ *of surfaces with irreducible free characteristic curves of genus* $p \geqq 3$, *and of freedom* $r \geqq 3p + 6$, *is birationally equivalent to a threefold generated by a pencil of rational surfaces, on which the characteristic curves of* $|F|$ *are mapped by unisecants to the pencil. If the system* $|F|$ *is simple, then this threefold is planar.*

Chapter V.
The Adjunction Problem.

1. Introduction. In the theory of surfaces there is a complex of results concerning the rational and scrollar types which presents a picture remarkable for its unity and simplicity. To begin with, we have CASTELNUOVO's theorem that any unirational surface is birational; then there is the CASTELNUOVO-HUMBERT theorem, according to which any surface containing an algebraic system, of index greater than unity, of rational curves, is rational. Next, any surface containing a pencil of rational curves is rational or scrollar, according as the pencil is rational or irrational; and moreover, in the latter case the pencil is unique. Again, such surfaces are characterised by the fact that their plurigenera are all zero; however, the only known proof of this result is long and difficult, involving not only CASTELNUOVO's theorem that a regular surface for which $p_2 = 0$ is rational but also the entire theory of the irregular surfaces of geometric genus zero.

An important result of different character, due to CASTELNUOVO and ENRIQUES [2], states that any surface on which the process of *successive adjunction*, applied to any curve system, terminates, is rational or scrollar; actually, for purposes of demonstration it is sufficient that the property in question should hold for any one system of general type: the proof consists in establishing the existence of a pencil of rational curves on the given surface. We then infer that, on any surface of plurigenera zero, the adjunction process always terminates; but a direct proof of this proposition has not yet been given.

All these results are classical; a different approach to the above problems, which proves useful in the theory of threefolds, is by way of the *anticanonical system**). We say that a surface possesses such a system if the virtual canonical system, reversed in sign, is effective. It is easily seen from first principles (cf. § 3) that the curves of this system are virtually elliptic, and that the surface has plurigenera zero, whence it must be either rational or elliptic scrollar; and quite simple considerations (ROTH [26]) suffice to exclude the latter possibility, at least when there are no exceptional curves of the first kind.

*) For this see SEVERI [13].

It is already known from counter-examples, and will further appear from others which follow, that most of the above theorems cannot be extended in their entirety to manifolds of higher dimension. Nevertheless, questions of the types we have described form a natural starting point for our investigations and in fact lead to many partial results of considerable interest. In the present chapter we shall be concerned mainly with the adjunction problem for threefolds and the questions that arise during the course of its examination.

If we approach this problem by seeking to imitate the procedure followed in the theory of surfaces, we at once find that the case of the threefolds is much more complex; this is due to the fact that, even on a completely regular threefold, the canonical system, and hence also the adjoint system, of a given surface, may present various features which have no analogue for a curve on a regular surface. We are therefore led to limit the problem in the first instance to certain categories of threefold in the hope that the material so obtained will guide us towards some result of general character.

2. Some varieties on which adjunction terminates. In the first place, let V_d ($d \geqq 3$) be any variety which can be mapped by an involution of some order on a variety V'_d; then if the adjunction process, applied to any system of hypersurfaces on V'_d, terminates, the same property must hold for V_d. This result is an immediate consequence of the transformation law for Jacobian systems – already alluded to several times – on two varieties in multiple correspondence. Hence, in particular, *on any unirational variety the adjunction process terminates*; for in such a case we may suppose V'_d to be a linear space.

Next, let V_d be a non-singular variety generated by a congruence of rational curves; we show that, on V_d, the adjunction process always terminates (Roth [32]). Suppose, first, that the congruence Γ has index unity; then we may assume that the generic curve $\mathfrak{C}$ of Γ is irreducible non-singular (IV, 4). We shall prove the result for the case $d = 3$; from this will follow the result for $d = 4$, and so on.

Assuming that $d = 3$, consider any surface R belonging to Γ; this is generated by a pencil of curves $\mathfrak{C}$ having the property that, if R' is an adjoint to R, $[\mathfrak{C}R'] = -2$. Hence the same result must hold for *any* curve $\mathfrak{C}$ of Γ. Now if A is any effective surface on V, its adjoint A' is given by the equivalence $A' \equiv A + R' - R$, from which it follows that $[\mathfrak{C}A'] = [\mathfrak{C}A] - 2$, and hence that the i-adjoint $A^{(i)}$ must be virtual for all sufficiently large values of i.

Proceeding next to the case $d = 4$, we first consider a threefold belonging to Γ, to which the previous result will apply, and then any threefold whatever; and similarly for $d = 5, 6, \ldots$.

Secondly, suppose that Γ has index $\nu > 1$; then, applying B. Segre's transformation (IV, 4), we obtain a variety V'_d simply generated by a congruence of rational curves, which is in $(\nu, 1)$ correspondence with V_d. And since, by what has been proved, the adjunction process terminates on V'_d, it must also terminate on V_d.

We observe that, *if the adjunction process on any variety V_d terminates, then the geometric genus and plurigenera of V_d must all be zero*, for otherwise the process would be non-terminating.

Turning now to the case $d=3$, we prove that, *if V_3 contains a birational congruence, of index unity, of rational curves, it is completely regular (of genus and plurigenera zero)*. We have already seen that the geometric genus P_g is zero; we now show that the superficial irregularity q_2 is likewise zero. Consider a net $|R|$ of rational surfaces in Γ, corresponding to a homaloidal net on the surface image of Γ; such nets certainly exist. Since the free characteristic curve of $|R|$ is irreducible, R is generic in the sense of CASTELNUOVO-ENRIQUES, and since R is regular, $q_2 = 0$.

Now, for any threefold, the character $P_g - P_a + q_2$ is non-negative (I, 10); hence, in the present case, $P_a \leqq 0$. Consider then two rational pencils $|A|$, $|B|$ of rational surfaces belonging to Γ and corresponding to two pencils (certainly existent) of unisecant rational curves on the surface image of Γ; we may always suppose these pencils so chosen that the curve (AB) is a single curve $\mathfrak{C}$, in which case it follows from a formula of B. SEGRE (II, 4) that the number t of stationary contacts of a surface A with a surface B is $t = 48(P_a + p_A + p_B + g_{AB})$, where p_A, p_B, g_{AB} denote respectively the arithmetic genera of A, B and the genus of the curve (AB).

Since, in the present case, $p_A = p_B = g_{AB} = 0$, while $t \geqq 0$, and $P_a \leqq 0$, we conclude that $P_a = 0$.

The first part of the proof shows incidentally that *any variety V_d which contains a birational congruence, of index unity, of rational curves, is superficially regular.*

We conclude this section with a first set of counter-examples, showing that *there exists a simply-infinite series of completely regular threefolds, on which the adjunction process terminates, which are birationally distinct and which are not unirational.*

Consider the product $V = \mathfrak{C} \times C$, where $\mathfrak{C}$ is a rational curve and C an irrational surface – or, what is the same thing, the point-cone projecting the surface C; since this contains a congruence, of index 1, of rational curves, it follows that, on V, the adjunction process terminates. It also follows from FANO's theorem (IV, 4) that V cannot be unirational. If we suppose that C is regular of geometric genus zero, then (I, 11) V is completely regular.

In order to construct a series of threefolds V we now appeal to the theory of the divisor (III, 4); from this we deduce that, if C has divisor σ, so also has V. Now GAETA [1] has established the existence of a family of surfaces, of which the ENRIQUES surface is the simplest, having characters $p_g = p_a = 0$, $p^{(1)} = 1$, $\sigma = 2^r$ $(r = 1, 2, 3, \ldots)$. Taking C to be each of these surfaces in turn, we obtain a simply-infinite series of threefolds which, since they have different divisors, must be birationally distinct from one another.

To obtain an example of a threefold with divisor unity which possesses the above properties, it would obviously suffice to find a surface with divisor unity and with invariants $p_g = p_a = 0$.

3. Threefolds which possess anticanonical systems. We now come to a second type of threefold on which adjunction terminates. Let V be a non-singular threefold, and $|F|$ any linear system of surfaces on V; then if the virtual impure canonical system $|F' - F| = |K|$, reversed in sign, is effective and has a non-zero order, we say that V possesses an anticanonical system, which we denote by $|A|$. We then have the following results (ROTH [23]):

I. *On V, the adjunction process always terminates.* For we have $|F^{(i)}| = |F + iK| = |F - iA|$, and the latter system must be virtual for all sufficiently large values of i.

II. *If the generic surface A is irreducible, it has geometric genus and plurigenera unity, and is free from exceptional curves* (that is, it has a pure canonical curve of order zero). For the adjoint A' is given by $A' \equiv A + K = 0$, and is therefore the null surface. Thus not only does it follow that A has geometric genus and plurigenera unity, but A cannot possess even an impure canonical system.

Assuming, then, that the generic A is irreducible, we have the following possibilities:

1. The system $|A|$ is a) free from base points, b) endowed with base points.

2. The generic A is a) regular, b) irregular.

3. V is a) superficially regular, b) superficially irregular.

4. V is a) tridimensionally regular, b) tridimensionally irregular.

As regards 1, simple base elements present no difficulty; but $|A|$ may *a priori* have multiple base elements of a kind that impose no conditions on the adjoint surfaces. If, when such elements exist, V could be transformed birationally in such a way as to eliminate them, then these would likewise present no difficulty; but since we do not know whether such a transformation is possible, we shall assume that condition 1a is satisfied.

We now show that

III. *If $|A|$ has freedom 3 at least, and A is regular, then V is tridimensionally regular.* For, by II., the grade n and genus p of any curve system on A must satisfy the relation $n = 2p - 2$; and since A is regular, the arithmetic genus of the surface is unity and hence, also, the curve system in question has freedom p. We may then calculate, by I, 4, the virtual characters of the system $|hA|$, where h is any integer; in particular, putting $h = -1$, we obtain the characters Ω_i of the system $|K|$, namely, $\Omega_0 = -2(p-1)$, $\Omega_1 = -3p+4$, $\Omega_2 = -p-2$. Inserting these values in SEVERI's relation, we deduce that the arithmetic genus of V is zero.

In one case to be considered later, $|A|$ is a net, with elliptic characteristic curves; but the above reasoning can still be applied.

If, in addition to being regular, A is generic, then of course V is superficially regular as well. We may add that, if A is irregular, then, by II., it must be a PICARD surface.

Conversely, *if V contains a general linear system of surfaces, free from multiple base points and exceptional curves, with geometric genus and plurigenera unity, then V possesses an anticanonical system.* For if $|A|$ is the system in question, its adjoint, which is well defined (I, 5), must be the null surface, so that $K = -A$. It now follows that, on V, the adjunction process terminates.

Evidently, if V contains a *net* of (irreducible) surfaces with pure canonical curves of order zero, its geometric genus P_g must be zero; for if $P_g > 0$, the characteristic system of the net, which is certainly effective, must be special, and in the present case this is impossible. On the other hand, V can contain a pencil, rational or irrational, of such surfaces; in that case the canonical and pluricanonical systems of V are compounded of this pencil (I, 12).

Although we shall be mainly concerned here with the case where V is completely regular, it is worth while pointing out that, in contrast with the theory of surfaces (§ 1), there exist various irregular types of threefold which possess anticanonical systems, as the following examples (ROTH [26]) show:

1. Suppose that V is the product of an elliptic curve and a rational surface which we assume to be free from exceptional curves of the first kind: we then see, by I, 11, that A is a PICARD surface which contains two pencils of elliptic curves. In this case V is planar.

2. Suppose instead that V is the product of a rational curve and a surface with an effective canonical curve of order zero; here the surface A consists of two PICARD surfaces belonging to a rational pencil, without base points.

3. We obtain a different type of threefold by making use of the fact (ENRIQUES, a) that a special PICARD surface can be mapped on a double cubic cylinder of S_3 with a branch curve consisting of four plane sections perpendicular to the generators. It follows that the threefold mapped on a double cylinder of S_4, whose base is a general cubic surface and whose branch surface consists of four prime sections perpendicular to the generators, contains a linear system, of freedom 3, of special PICARD

surfaces, whose characteristic system is compounded of elliptic curves of a congruence.

From now on we assume that the threefold V under consideration contains a linear system $|A|$ of *regular* surfaces which is subject to the assumptions made above; there are then three cases to examine:

I. If the system $|A|$, with curvilinear genus p, is completely irreducible, its freedom, by the RIEMANN-ROCH theorem, is $p+1$.

II. If the characteristic series of $|A|$ is compound, the characteristic curves must be hyperelliptic, and the series in question must be compounded of an involution I_2; here again the freedom of $|A|$ is $p+1$.

III. If the characteristic system of $|A|$ is compound, it must be composed of a congruence, of index unity, of elliptic curves; for, since the grade of the system is zero, its virtual genus must be unity (p. 79). The generic surface A contains a linear pencil of such curves; hence the congruence is rational, since it may be generated by a linear pencil of $|A|$, each surface of which is itself generated by a linear pencil of the curves in question.

Case I will be discussed in detail in the following sections; here we merely remark that since $p \geqq 3$ (for, if $p=2$, we are led to case II), we may use the system $|A|$ to map V birationally on a completely regular threefold W^{2p-2} of S_{p+1}, whose generic curve section is canonical of genus p: thus W^{2p-2} cannot possess multiple surfaces, though it may have multiple curves or points. We call W^{2p-2} a FANO *threefold*.

In case II, we may map V on a double threefold W^{p-1} of S_{p+1}, with rational curve sections; this is plane-ruled, except possibly when $p=5$, in which case it may be a VERONESE cone (IV, 9). The branch surface is a F^{2p+2} which in general is irreducible non-singular. Since W^{p-1} is birational, we may, if convenient, pass to a representation on a double S_3.

Suppose first that $p=5$, and that W^4 is a VERONESE cone; then the generic A contains a net of curves of grade and genus 2, and the branch surface F^{12} is the complete intersection of W with a cubic primal not passing through the vertex O of W^4 (for otherwise the generic section through O would not represent a surface of genera unity). Since W^4 contains a net of quadric cones, with vertices at O, any two of which have a common generator, it follows that V contains a net of rational surfaces of the fourth family (IV, 2), any two of which meet in an elliptic curve. Now this is a case where our previous method of establishing unirationality cannot be applied: hence nothing further can be asserted, on this basis at least.

Next, suppose that W^{p-1} is plane-ruled. Since the generic prime section R^{p-1} contains a rational pencil of lines, the corresponding surface A contains a rational pencil of curves without base points; and since these curves must be elliptic, the branch surface F^{2p+2} must be cut on W^{p-1}

by a quartic primal, residual to $2(p-3)$ generating planes. This surface meets the generating planes in quartics, which in general are non-singular.

Taking a prime section, we find that F^{2p+2} has sectional genus 9. The prime section of F^{2p+2} residual to the quartic in which it meets a given generating plane π is a curve meeting the quartic in four points, and thus has genus 3. From this it follows that V cannot exist for $p > 4$. For suppose, if possible, that $p \geqq 5$; then we could draw a space S_5 through π and any other generating plane π'; and since both π and π' meet F^{2p+2} in non-singular quartics, the residual section would have to consist of $2(p-3)$ lines, so that F^{2p+2} must be a scroll lying in the S_5 containing π and π'.

We now show that, *for* $p = 3, 4$, *V is unirational.* For the generating planes of W^{p-1} map a rational pencil of surfaces $|A_1|$ which, by IV, 2, are unirational in $K(P)$, while the residual scrolls R^{p-1} map rational surfaces A_2, meeting each A_1 in elliptic curves. In order to obtain a unirational representation for V, we have only to fix, on a particular surface A_2, a rational curve which is plurisecant to the pencil $|A_1|$; and this can be effected in an infinity of ways.

Consider next case III, in which the characteristic system of $|A|$ is compounded of a congruence Γ (rational and of index 1) of elliptic curves $\mathfrak{C}$. Supposing that the generic characteristic curve consists of n ($\geqq 1$) curves $\mathfrak{C}$, we can, as in IV, 10, represent V on an n-ple cone W ($n \geqq 1$) which, since V is superficially regular, is normal in S_{n+2} and has rational curve sections of order n. Hence W is either generated by ∞^1 planes or else is a VERONESE cone ($n = 4$).

If $n \geqq 2$, and W is plane-ruled, the system $|A|$ therefore contains two subsystems $|A_1|$ and $|A_2|$, say, of irreducible non-singular surfaces, corresponding respectively to the generating planes and residual prime sections of W, and hence of respective freedoms 1 and $n-1$. The generic A_1 meets the generic A_2 in an elliptic curve $\mathfrak{C}$, and each such surface contains a rational pencil of curves $\mathfrak{C}$. If instead W is a VERONESE cone, $|A|$ contains a net $|A_1|$ such that $A \equiv 2A_1$, with characteristic curve $\mathfrak{C}$, and where A_1 likewise contains a pencil $|\mathfrak{C}|$.

We now show that *each of the surfaces* A_1 *and* A_2 *is rational.* First, suppose that W is planar; then, from the relation $A \equiv A_1 + A_2$, where, by hypothesis, $A' = 0$, it follows that neither A_1 nor A_2 can possess an effective canonical system, so that each surface has geometric genus zero. Again, since the virtual arithmetic genus of the composite surface $A_1 + A_2$ is unity, while the component surfaces meet in an elliptic curve, and since also neither surface can have arithmetic genus greater than zero, it follows that each has arithmetic genus zero. In order to prove that each surface is rational, it now suffices, by CASTELNUOVO's theorem (§ 1), to show that each has bigenus zero. This is a consequence of the formulae (Enriques, a) for the plurigenera of a surface which contains a rational pencil of elliptic curves, from which it appears that if such a surface has invariants $p_g = p_a = 0$, $p_2 > 0$, the pencil must possess a certain number of multiple elliptic curves. On the other hand, we know, from the same formulae, that the pencil $|\mathfrak{C}|$ on A cannot contain multiple members, whence Γ cannot possess such curves. Precisely similar reasoning shows that, if W is a Veronese cone, the surface A_1 is rational.

It is possible for the surface A_2 to break up into A_1 and a residual component A_3 which, by what has just been proved, must in general be elliptic scrollar; but no further degeneration is possible, since a surface for which $p_a < -1$ is scrollar of genus greater than unity and so contains no elliptic curves.

This last remark shows that the only possible cases are $n = 1, 2, 3, 4$. For $n = 3$, we have the degeneration $A = 2A_1 + A_3$, where A_3 corresponds to the directrix plane of W^3; this surface may be rational, provided that A_1 meets A_3 in a rational curve: otherwise it must be elliptic scrollar. For $n = 4$, we must evidently take W^4 to be a cone which possesses a directrix plane, since otherwise there can be no permissible degeneration. But when $n = 5$, even if we assume the existence of such a plane, we obtain impermissible degenerations.

Finally, we consider the question of unirationality. In the case $n = 2$, V contains two pencils $|A_1|$, $|A_2|$ of rational surfaces; hence, by IV, 4, if either A_1 or A_2 belongs to one of the first three families, V is unirational. Similar results are found in the cases $n = 3, 4$. When W is planar, we use the fact that V contains a pencil $|A_1|$ of rational surfaces and at least a net $|A_2|$ of such surfaces; when W is a Veronese cone, V contains a net $|A_1|$ of rational surfaces from which we may extract a pencil and a single surface not belonging to the pencil; hence the same conclusion holds in these cases.

When, however, all the surfaces in question belong to the fourth family, the above methods fail; in that case, by IV, 2, we can map V on a double S_3 with branch surface F^{2m} ($m \geqq 3$) endowed with two consecutive $(2m-3)$-fold points whose join is $(2m-6)$-fold on F^{2m}; and, possibly, additional singularities, which may, in certain favourable instances, render V unirational or birational.

4. The Fano threefolds. Let W^{2p-2} be a completely regular threefold whose generic curve section $\mathfrak{C}$ is canonical of genus p ($p \geqq 3$), and whose generic prime section F^{2p-2} is a regular surface of genera and plurigenera unity which is free from exceptional curves; then W^{2p-2} is normal in S_{p+1} and, incidentally, is not a cone. A general investigation and classification of W^{2p-2} has been given in Fano [12], after a preliminary study, to which we shall refer later, of various particular cases on which some of the main conclusions rest. The following survey of Fano's work, while perforce incomplete, will indicate sufficiently the variety of applications of the theories of surfaces and threefolds which is to be found in it.

The essential technique is an exploitation of the properties of canonical curves and hence of surfaces with such curves as their prime sections; these are then transmitted to the threefolds under investigation. To begin with, FANO shows that W^{2p-2} cannot be generated, simply or multiply, by a congruence of lines: this result will be useful later. Another important result, of elementary character, concerns the effect of projecting the given threefold from an appropriate vertex: thus W^{2p-2} projects from a line of itself into a W^{2p-6}, also of canonical curve sections, on which the line is mapped by a normal cubic scroll; while W^{2p-2} projects from a conic into a W^{2p-8} of canonical curve sections which contains a rational quartic scroll, image of the conic in question. Again, the projection of W^{2p-2} from the tangent space at a general point is a W^{2p-10} of canonical curve sections, on which the neighbourhood of the vertex is mapped by a VERONESE surface.

The next group of properties arises from the consideration of two series which are mutually residual with respect to the canonical series: we readily prove that, if a prime section of F^{2p-2} breaks up into two irreducible components, each must be normal non-special, and the complete system defined by each is cut out by primes through the other; also each such system cuts a complete series on the generic prime section of F^{2p-2}. Further, it appears that any irreducible non-normal curve partially contained in a prime section of F^{2p-2} must be residual to a reducible system which either has an irreducible variable component and a rational fixed part or else is compounded of elliptic curves with (possibly) a rational fixed part.

From this proposition we deduce another concerning W^{2p-2} by the following considerations. Let Φ be a pencil of sections F passing through a given irreducible curve section $\mathfrak{C}$ of W^{2p-2}; and suppose that on each F there is a well defined linear system $|\mathfrak{D}|$ of normal curves of order n and genus π which is partially contained in the prime sections of F and which therefore cuts a complete series g_n^π on $\mathfrak{C}$. This series cannot vary with F, for otherwise it would describe an ∞^1 rational system (in correspondence with the surfaces of Φ, or with an involution in Φ) and so the sets of these ∞^1 series which contain π assigned points of $\mathfrak{C}$ would form an ∞^1 rational system of disequivalent sets, which is impossible. If then we consider, on the surfaces F of Φ, the curves $\mathfrak{D}$ which pass through π assigned points of $\mathfrak{C}$, the locus of such curves is a surface G of order n (since it meets each F in a curve $\mathfrak{D}$) lying in a space of dimension superior by unity to that of the space containing $\mathfrak{D}$ and hence contained partially in the system of prime sections of W^{2p-2}.

And the residual system $|F - G|$ will cut a complete system of curves on these prime sections.

Suppose instead that, on each surface F, there exist two or more systems such as $|\mathfrak{D}|$, with the same characters; then it can be shown that, as F varies in Φ, each system $|\mathfrak{D}|$ describes a separate system: this result is a consequence of the fact that, on any surface of genera and plurigenera unity, the division of linear systems, when possible, can be effected in only one way (III, 4). In conclusion, then:

Any complete linear system of irreducible normal curves on the generic prime section F of W^{2p-2} and contained in the prime sections of F is cut on F by a linear system of surfaces of the same order lying on W^{2p-2}. If the system $|F|$ is the sum of two irreducible systems, each of these systems cuts on F a complete linear system of normal curves.

A second fundamental result, deduced from the same considerations, is as follows:

If the system $|F|$ of prime sections of W^{2p-2} is the sum of two irreducible systems neither of which is fundamental) for $|F|$, then the surfaces of each system are rational and the common curves of two surfaces of the two systems are elliptic.*

If n, π are the order and sectional genus of either system $|G|$, say, it follows from the preliminary result concerning series that $n > 2\pi - 2$, so that the surfaces of the system are either rational or scrollar. Now they cannot be irrational scrolls; for they would then cut on F a linear system $|\mathfrak{D}|$ of curves of genus 1 at least and so of freedom 1 at least, and the system would itself have freedom 1 at least, so that W^{2p-2} would contain ∞^2 lines. Hence, if $|\mathfrak{D}|$ has freedom (and therefore genus) zero or unity, G must be rational. Suppose, if possible, that G is irrational scrollar; then, since W^{2p-2} is superficially regular, $|G|$ must have reducible characteristic curves which belong to a congruence of index 1 and which have order $k > 1$; hence the system $|\mathfrak{D}|$ on F must belong to the involution I_k cut on F by the congruence, so that the characteristic series of $|\mathfrak{D}|$, which is the canonical of a curve $\mathfrak{D}$, must be compounded of I_k; and this is possible only if $k = 2$, and if $\mathfrak{D}$ is hyperelliptic. It follows that G has hyperelliptic curve sections (of genus greater than unity) and since the surface is not ruled, it is rational. Finally, since both G and $F - G$ are rational, the two surfaces meet in an elliptic curve.

5. Classification of the FANO threefolds. The classification of the varieties W^{2p-2} depends in the first instance on the nature of F; we say that F is of the first species**) if it contains only complete intersections, in which case we can show that W^{2p-2}, likewise termed of the first species, contains only complete intersections. We say that F is of the second

*) Thus the case of an isolated double point of W^{2p-2} is excluded.

**) We may point out that a rigorous proof of the existence theorem for these surfaces has yet to be given.

species if, on F, a base is provided by a submultiple of a prime section; in that case W^{2p-2} will be endowed with an analogous property. All other types are said to be of the third species; among them are those whose prime sections contain a g_3^1, in which case it can be shown that F contains a pencil of plane cubics and that W^{2p-2} possesses a pencil of cubic surfaces which are in general non-singular. These definitions differ from FANO's in that FANO includes among the first or second species any surface which belongs to a subclass of either.

It is important to observe that a *single* condition is imposed on a surface of the first species if it is to contain a line or a conic (ENRIQUES, a). It follows from this that every threefold of the first species contains a scroll of lines and a congruence of conics; and it is easily shown (FANO [11]) that, if a scroll – which must be a complete intersection – is cut on the threefold by a primal of order k, then the generic generator of the scroll is met by $k+1$ others. A like conclusion applies to any component of a reducible scroll, if there is any such; also if a threefold of the third species contains a scroll or scrolls, each generator is met by a certain number of other generators; these give rise to nodes on the projection of the threefold from the line in question. These threefolds, and others which do not contain scrolls, may possess a finite number of planes.

Consider first those threefolds which contain scrolls (possibly reducible); denoting by m the maximum number of lines, for a given value of p, which meet an assigned line of a threefold W^{2p-2}, we see that this character is certainly augmented by projection from this line, so that, when p is increased by 2, m must be diminished. Now it is found from a study (FANO [11]) of particular cases that, for $p = 7, 8$, $m = 8, 6$; hence, for $p > 7 + 2.8$, i. e. $p > 23$, if there exist threefolds containing scrolls, for such scrolls the character m must be zero, which means that they are projections of FANO threefolds in higher space which contain no lines other than those lying in isolated planes.

We shall see that all the varieties W^{2p-2} which contain scrolls are birational for $p > 10$; and that *all* the FANO threefolds are birational for $p > 10$, except possibly when $p = 13$.

The threefolds of the first species, though few in number, are specially interesting for several reasons; leaving them for detailed consideration later, we pass to the threefolds of the second species. For these the prime sections F are of the form $F = kG$, where G is irreducible, and by § 4, rational; and since G is a base on W^{2p-2}, the system $|G|$ must be ∞^3 at least, with simple characteristic system. The characteristic curves are elliptic when $k=2$, and rational when $k > 2$.

If $k > 2$, the system $|G|$ must be simple and hence W^{2p-2} is birational (IV, 9), with a representation on S_3 in which $|G|$ is mapped by one of the systems of surfaces in IV, 9; however, it is found that the sole types having a multiple of the required character are the system of planes and the system of quadrics through a base conic, for which $k = 4, 3$ respectively.

If instead the characteristic curves of $|G|$ are elliptic, W^{2p-2} is certainly birational if $|G|$ has grade $n > 3$; when $n = 3$, W^{2p-2} is birationally equivalent to a cubic primal of S_4 and, when $n = 2$, to a double S_3 with quartic branch surface. Since $k = 2$, in all these cases $|F|$ has grade $8n$ $(= 2p - 2)$; and since, as we know, $n \leqq 9$, it follows that $2p - 2 \leqq 72$, whence $p \leqq 37$. In point of fact the threefold corresponding to the value $p = 37$ is one of maximum order; evidently it is mapped on the cone which projects a DEL PEZZO novenic surface from an external point by the quadric sections of this cone.

We may refer to FANO for the rest of this classification, and turn to the threefolds of the third species or, more precisely, those which are not subclasses of the first or the second species, which are the only types examined by FANO in detail. For these we have a relation of the form $F = \Sigma k_i G_i$, where $k_i \geqq 2$, except in the case where W^{2p-2} contains a pencil of cubic surfaces; otherwise, if some coefficient k_i is unity, we obtain one of the above-mentioned subclasses.

Consider then the system $\Sigma \left[\frac{k_i}{2}\right] G_i$, where $\left[\frac{k_i}{2}\right]$ denotes the integral part of $\frac{k_i}{2}$; this system is irreducible and consists of rational surfaces not belonging to a congruence of curves. The system $\Sigma 2 \left[\frac{k_i}{2}\right] G_i$ is obviously contained in $|F|$, and we may suppose that it is partially contained therein, i. e. that one at least of the numbers k_i is odd $(\geqq 3)$, since otherwise W^{2p-2} would belong to a subclass of the second species. It follows that the system $\Sigma 2 \left[\frac{k_i}{2}\right] G_i$ consists of rational surfaces; hence the system $\Sigma \left[\frac{k_i}{2}\right] G_i$ has rational characteristic curves and is therefore simple; and the system residual to $\Sigma 2 \left[\frac{k_i}{2}\right] G_i$ with respect to $|F|$ is the sum of those systems G_i for which the coefficients k_i are odd, each system being counted once. Hence this residual system will be contained in $\Sigma \left[\frac{k_i}{2}\right] G_i$, and we may suppose it to be partially contained therein, since otherwise all the numbers k_i would be equal to 3, and we should once more have a subclass of the second species. It follows that we can map W^{2p-2} on S_3 by means of the system $\Sigma \left[\frac{k_i}{2}\right] G_i$; and the system $|F|$ will then correspond to one more ample than $\Sigma 2 \left[\frac{k_i}{2}\right] G_i$ and less ample than $\Sigma 3 \left[\frac{k_i}{2}\right] G_i$.

In these cases, then, the representative system may be referred to that of the prime sections of a VERONESE cone or a rational normal planar threefold which, as is easily seen, can be at most of order 8. In this last case the threefold V_3^8 must be a cone with quadric directrix; the system $|F|$ is mapped by cubic sections of V_3^8 residual to six generating planes, and gives the second type of FANO threefold of maximum order 72; this has two 8-ple points whose join is a double line.

Finally, we have to determine the threefolds W^{2p-2} which contain pencils of cubic surfaces; these lie on rational normal fourfolds V_4^{p-2} generated by the ambient spaces of the cubic surfaces, and it may be shown that they are obtainable as intersections of V_4^{p-2} with cubic primals passing through $p-4$ generating spaces. A detailed analysis (FANO [12]) shows that, for the essentially new types, $p \leqq 10$; for higher values of p, the threefolds obtained are special cases of those considered above.

6. Conditions for unirationality or birationality. The examination of particular cases, corresponding to values $p \leqq 7$ (FANO [11, 12]) is not only interesting for its own sake but is a necessary preliminary to what follows. To this end we first determine (cf. ENRIQUES-CHISINI, a) all the canonical curves, both general and particular, of genera $p \leqq 7$, and from them we pass to the corresponding surfaces F and thence to the threefolds W^{2p-2}. By quite elementary means, which will be illustrated shortly, we may then show that, *for* $4 \leqq p \leqq 7$, W^{2p-2} *is in all cases unirational or birational.*

We now prove the following lemma: *the threefolds* W^{2p-2} *which contain a plane are all unirational.*

The result, for $p = 3$, is immediate; we then have a quartic primal W^4 containing a plane π and a pencil of cubic surfaces residual to it. This pencil has base points in π; and it will be recalled (IV, 3) that any cubic surface is unirational in $K(P)$.

Supposing now that $p \geqq 4$, we observe that, through the plane π in question pass ∞^{p-2} primes, cutting W^{2p-2} residually in rational surfaces G, of sectional genus $p-2$, which meet π in cubics. If the system $|G|$ is completely irreducible and of freedom 4 at least, it follows from FANO's theorem (IV, 11) that W^{2p-2} is either birational or birationally equivalent to a cubic primal of S_4; the excluded case $p = 5$ has already been dealt with in the preliminary investigation above.

We have next to examine the case where the characteristic system of $|G|$ is compounded – as certainly happens for $p = 4$ – of a congruence. Now the characteristic curves are the intersections of pairs of primes, residual to π, and are therefore of order $2p-6$; and since each surface G contains an ∞^{p-3} system of such curves, the curves of the congruence have order $(2p-6)/(p-3)$ at most, and are thus conics. On each G there is a rational pencil of these conics; the planes of the conics meet π

in lines, so that the conics themselves are bisecant to π, from which it follows that W^{2p-2} is unirational and representable on an I_2.

Lastly, we have to consider the possibility that the characteristic series of $|G|$ is compound; in this case, appealing to the result of IV, 11, we reach the conclusion stated. But it is simpler to use first principles, as in FANO [12]. Let k ($\geqq 2$) denote the number of points in which a characteristic curve $\mathfrak{D}$ of $|G|$ meets π; then $\mathfrak{D}$ has genus $p-2-k$, and on each G the linear system of such curves has a characteristic series of order $2p-6-k$ and freedom $p-4$. If this series is compounded of an involution I_m, we must have $p-4 \leqq (2p-6-k)/m$. Rewriting this inequality in the form $(p-4)(m-2) \leqq 2-k$, we see that, for $p > 4$, the only solution is $m = k = 2$. Thus we find that W^{2p-2} lies on a variety V_4^{p-2}, locus of the ∞^3 lines, which are incident to π, joining point pairs of I_2; it is easily seen that W^{2p-2} must contain a pencil of cubic surfaces, with rational unisecant, and hence that it is unirational.

From the above lemma we deduce the theorem: *if $p > 3$, W^{2p-2} is unirational.*

Supposing, as we may, that $p > 7$, we project W^{2p-2} from the tangent space at the general point, thereby obtaining (§ 4) a W^{2p-2} of S_{p-3} with, canonical curve sections of genus $p-4$, containing a VERONESE surface F^4. If, then, p lies in the range $7 < p < 12$, we have one of the threefolds already examined (possibly) particularised to the extent of containing a VERONESE surface. It is easily seen that the presence of this surface does not affect the previous arguments, though in some cases it may be used to simplify them.

For values of $p \geqq 12$, we make a second projection, from a conic of F^4, so obtaining a W^{2p-16} of S_{p-6}, with canonical curve sections of genus $p-7$, containing a plane which is the projection of F^4. By the result just proved, this threefold is unirational.

The above lemma, and consequently the deductions made from it, can be improved by a more detailed analysis which, however, partly depends on FANO's classification of threefolds of the third species.

Using these results we can prove that *any W^{2p-2} which contains a plane but not a pencil of cubic surfaces is birational for $p > 6$, and that, if W^{2p-2} contains such a pencil, it is birational for $p > 7$.* From this we deduce the theorem: *any W^{2p-2} which contains a scroll is birational for $p > 10$.* For the projection of W^{2p-2} from a generator of the scroll is a W^{2p-6} of S_{p-1} containing a cubic scroll R^3; and the projection of W^{2p-6} from a generator of R^3 is a W^{2p-10} of S_{p-3} containing a plane which is the image of R^3.

We may also prove that, *if W^{2p-2} contains no lines save (possibly) those which lie in a number of isolated planes, it is birational for $p > 13$.* This result follows from the lemma by projecting W^{2p-2} from the tangent space at a general point, so obtaining, as before, a W^{2p-10} containing a

VERONESE surface F^4; and then projecting again from a conic of F^4, thereby obtaining a W^{2p-16} which, by what has been proved, is birational if*) $p-7>6$, i. e. if $p>13$.

We are now in a position to conclude the investigation of the anticanonical systems proposed in § 3 with the following general statement (ROTH [23]):

A completely regular non-singular threefold V which possesses an anticanonical system $|A|$ has the property that, on V, the adjunction process always terminates. If the generic surface A is irreducible, and if $|A|$ is free from base points, there are three possibilities to consider:

I. The system $|A|$ is completely irreducible; then the curvilinear genus p must satisfy the inequality $p \leqq 37$; also, if $p>3$, V is unirational and, if $p>10$, V is either birational or birationally equivalent to a cubic primal of S_4.

II. The characteristic series of $|A|$ is compound; then the characteristic curves (of genus p) are hyperelliptic. The only admissible values of p are $p=2, 3, 4, 5$: for $p=3, 4$, V is unirational, while for $p=2, 5$, it is representable on a double S_3 with sextic branch surface.

III. The characteristic system of $|A|$ is compound, in which case the characteristic curve is composed of n elliptic curves of a rational congruence of index unity. The only admissible values of n are $1, 2, 3, 4$; for $n=2, 3, 4$, V is either unirational or representable on a double S_3 of which the branch surface is a $F^{2m} (m \geqq 3)$ endowed with two consecutive $(2m-3)$-ple points, joined by a $(2m-6)$-ple line; and, possibly, additional singularities.

7. Threefolds of the first species. Suppose now that W^{2p-2} is of the first species; we then find that, for $p=3, 4, 5, 6, 8$, respectively W^{2p-2} is a quartic primal, the complete intersection of a quadric and a cubic primal of S_5, the complete intersection of three quadrics of S_6, a section of the V_5^{10} in which a quadric meets the Grassmannian of the lines of S_4, and a section of the Grassmannian of the lines of S_5. For each such value of p, the order of the scroll on W^{2p-2} and the corresponding character m are known, partly from the work of earlier writers and partly by application of the method of projection from a line, combined with those results.

We have already shown in IV, 8 that W^6 is unirational, and have remarked that it is representable on an I_{36}; and we have shown in IV, 5 that W^8 is unirational and representable on an I_4. Next, projecting W^{10} from a line, we obtain a W^6 (no longer of the first species) which contains a cubic scroll; it can then be shown (FANO [11]) that the order of the corresponding involution can be lowered from 36 to 6. Finally,

*) Actually as a result of the classification it can be asserted that W^{2p-2} is birational for $p>10$, save possibly for $p=13$, in which case W^{2p-2} is birationally equivalent to a cubic primal of S_4.

the projection of W^{14} from a line is a W^{10} containing a cubic scroll R^3; and it can be proved (FANO [17]) that the ∞^2 primes through R^3 meet W^{10} residually in surfaces of sectional genus 2 which meet in pairs in conics bisecant to R^3. Since such conics form a rational congruence of index unity, it follows that W^{14} is unirational and representable on an I_2.

Consider next the case $p = 7$; here the corresponding W^{12} is birational, since it projects from a line into a W^8 containing a scroll R^3 through which pass ∞^1 primes meeting W^8 residually in DEL PEZZO quintics, whence the result follows from IV, 10. Alternatively, we may project W^{12} from a tangent space, thus obtaining a primal W^4 which contains a projected VERONESE surface; and since the latter possesses one apparent triple point, if follows that W^4 can be projected birationally on to a prime (FANO [15]). The cases $p = 9, 10$ are dealt with in the latter work; thus W^{16} projects from a line into a W^{12} containing a scroll R^3, through which pass ∞^3 primes meeting W^{12} residually in surfaces which, as is readily seen, form a homaloidal web, so that W^{16} is birational. Again, W^{18} projects from a line into a W^{14} containing a scroll R^3, through which pass ∞^4 primes cutting W^{14} residually in a system of surfaces which, as is easily shown, has rational characteristic curves and grade 2, so that it can be mapped on a quadric of S_4. Thus W^{18} is likewise birational. From these projections we may determine the orders of the scrolls, and the values of the character m, for W^{16} and W^{18}; and then, using the method of successive projection from lines, together with the previous results, one can show (ROTH [19]) that, for $p > 10$, threefolds of the first species are non-existent.

It is now possible to give a simpler and more complete solution of the adjunction problem for the completely regular threefolds of base number unity (ROTH [19]). Let V be a completely regular threefold having the surface B for base; then B is necessarily regular, and the complete system $|B|$ defined by B must be free from base points and its characteristic system must be simple. We first show that *a necessary and sufficient condition that the adjunction process should terminate on V is that the system $|B'|$ should be null or virtual.* For let $B' = hB$, where h is any integer, including zero; then, if n is any positive integer,

$$(nB)' \equiv (n-1)\,B + B' \equiv (n+h-1)\,B\,.$$

Thus the adjunction process terminates for the system $|nB|$ if, and only if, $h \leqq 0$.

Suppose first that B' is virtual, in which case a virtual canonical surface K of V is of the form $K = -kB$ $(k \geqq 2)$. Denoting by n, p the virtual grade and curvilinear genus of $|B|$, we can calculate, by I, 4, the corresponding characters Ω_0, Ω_1 of $|K|$; substituting these results in the relation (I, 2)

$$2\,\Omega_1 - 2 = 3\,\Omega_0\,,$$

we obtain

$$2(p-1) = (2-k)\,n\,.$$

Since $n > 0$, it follows that $p = 0$, or $p = 1$ (with $k = 2$). And, when $p = 0$, we see that this equation has precisely two solutions, namely $n = 1$, $k = 4$; and $n = 2$, $k = 3$. We have here an arithmetical proof of the fact that the only threefolds with rational curve sections and base number unity are the space S_3 and the general quadric of S_4.

In the case $p = 1$, V has on it a linear system, of freedom 3 at least, of rational surfaces with elliptic characteristic curves, and is therefore birational or unirational.

Next, supposing that B' is the null surface and that the system $|B|$ is simple, we see that V must be a Fano threefold of the first species, so that no further discussion is necessary. If instead $|B|$ has a compound characteristic series, its characteristic curves must be hyperelliptic, in which case V can be mapped on a double threefold W with rational characteristic curves, containing only complete intersections. Hence W is either a double S_3 with sextic branch surface ($p = 2$) or a double quadric of S_4 with octavic branch surface ($p = 3$). The former type is presumably not unirational; the latter is unirational, as may be seen from the method of IV, 8.

We conclude our account of the Fano threefolds with a remark, due to Fano, concerning a certain contrast between the invariant theories of surfaces and threefolds. We know that the absolute linear genus $p^{(1)}$ of a rational surface has the value 10, and that the number $p^{(1)} - 1$ represents the maximum dimension of a linear system of elliptic curves on the surface in question; in fact such a system is given by that of all plane cubics. Now in the theory of threefolds the analogous invariant is the number $-\Omega_2$; its value for a space S_3 is 35, and in point of fact the system of quartic surfaces of S_3 has dimension 34. But this number is *not* the corresponding maximum for birational threefolds: for example, the cone which projects a Del Pezzo novenic surface has $-\Omega_2 = 39$. Again, there exist in S_3 linear systems of regular surfaces of genera and plurigenera unity which have dimension greater than 34; for instance, systems which map Fano threefolds of high order: thus quadric sections of the above-mentioned Del Pezzo cone give a system of the required type having freedom 38. In view of these facts Fano has proposed that a suitable absolute invariant for a threefold would be the maximum dimension of the systems of regular surfaces of genera and plurigenera unity which it contains; in the present state of the theory, however, the calculation of such an invariant usually presents insuperable difficulties.

The varieties W_d^{2p-2} $(d > 3)$ with canonical curve sections which contain only complete intersections with primals have been considered in ROTH [16], where it is shown that such varieties are birational for $p \geqq 7$, $d \geqq 4$; $p = 5, d \geqq 5$; and for $p = 6$, $d \geqq 5$. The representations on S_4 of the birational types for which $d = 4$ have been determined in ROTH [17], from which it follows that such fourfolds exist if, and only if, $p \leqq 10$.

A particular fourfold of the third species, namely the W_4^{22} which maps the chords of a rational normal quartic scroll on the Grassmannian of the lines of S_5, has been examined by FANO [18].

8. Threefolds whose prime sections are ENRIQUES surfaces. Consider a completely regular non-conical threefold whose prime sections are ENRIQUES surfaces; assuming that the generic prime section is non-singular and free from exceptional curves, we know that its order must be $2p - 2$, where p $(\geqq 4)$ is the sectional genus. Thus the threefold V^{2p-2} in question is normal in S_p; we shall show that it is intimately connected with the FANO threefolds – a circumstance first noticed incidentally by GODEAUX [1]: in fact we prove that *V^{2p-2} contains an ∞^{p-1} linear system of surfaces of genera and plurigenera unity, of curvilinear genus $p - 2$ and grade $2p - 6$.*

To this end we remark that the generic prime section F of V^{2p-2} contains, in addition to the system $|\mathfrak{C}|$ of its prime sections, a system $|\mathfrak{D}|$ of the same characters, which has the properties that $\mathfrak{C}$ and $\mathfrak{D}$ are mutually adjoint, and that $|2\,\mathfrak{C}| \equiv |2\,\mathfrak{D}|$; the curves $\mathfrak{D}$ are canonical of genus p, since they lie in spaces S_{p-1}. Now fix $p - 1$ general points on a curve $\mathfrak{C}$, and consider on the surfaces F of the pencil containing $\mathfrak{C}$ the unique curves $\mathfrak{D}$ through these points; the locus of such curves $\mathfrak{D}$ is a surface G having these curves for prime sections. Evidently G is regular of genera and plurigenera unity; it is variable in an ∞^{p-1} system; and we readily deduce that this system has curvilinear genus $p - 2$ and grade $2p - 6$. It can be shown (FANO [13]) that, if $p \geqq 5$, the system $|G|$ is completely irreducible; when $p = 4$, however, the characteristic series of $|G|$ is compounded of an involution I_2.

Assuming, then, that $p \geqq 5$, we can map the system $|G|$ on the prime sections of a FANO threefold W^{2p-6}. Now since, on any surface F, we have $2\,\mathfrak{C} \equiv 2\,\mathfrak{D}$, and since $|2F|$ and $|2G|$ cut equivalent curve systems on F, it follows that the surfaces $2F$ and $2G$ are either equivalent or they differ by fundamental surfaces of $|F|$, i. e. multiple points of V^{2p-2}; on W^{2p-6}, these will be represented by actual surfaces. And since the grades of $|2F|$, $|2G|$ are $8(2p - 2)$ and $8(2p - 6)$ respectively, the difference, 32, between these two numbers in the sum of the multiplicities of the singular points of V^{2p-2}: the system $|2G|$ will be cut on V^{2p-2} by quadrics passing through these points.

On W^{2p-6} the systems $|f|$, $|g|$ which map $|F|$ and $|G|$ respectively have orders $2p - 2$, $2p - 6$, while the systems $|2f|$, $|2g|$ have orders $4p - 4$,

$4p-12$; hence the surfaces $2f$, $2g$ differ by a composite surface of order 8. We shall show that the latter consists of 8 planes.

First, we observe that the addition of this composite surface to $2g$ cannot alter the genus of the surface; for the system $|f|$ is formed of regular surfaces of genus zero, meeting in curves of genus p, while the system $|g|$ is formed of regular surfaces of genus unity, meeting in curves of genus $p-2$, from which it follows that the surfaces $2f$, $2g$ are both regular of genus p. Hence the surfaces which are fundamental for $|f|$ must be met by the surfaces $2g$, i. e. by quadrics of the ambient space, in curves of genus zero; and this is possible only if the composite fundamental surface consists of 8 planes. On V^{2p-2} each of these planes is mapped by a quadruple point.

No two planes can have a common line, for in that case they would constitute a reducible quadric which, when added to $2g$, would increase the genus of the surface. Two planes can, however, meet in a point, which will then be double for W^{2p-6}.

In FANO [13] the discussion is confined to the general case in which no three planes have a common point; it is found that there are precisely four types of threefold, corresponding to the values $p = 6, 7, 9, 13$; each of these threefolds is birational, and each of the corresponding FANO threefolds is mapped on S_3 by a system of quartic surfaces. The representations of V^{2p-2} on S_3 are also obtained; and thus, in the general case at least, FANO has solved the problem of determining all simple linear systems of ENRIQUES surfaces in S_3 which have freedom $r \geqq 5$.

In the case $p = 4$, the system $|G|$ belongs to an involution I_2, so that the previous mapping of V^{2p-2} on W^{2p-6} is no longer possible. The threefold V^6, which we call the ENRIQUES *primal,* evidently possesses six double planes, which are the intersections in pairs of four primes, and four triple lines, intersections in threes of those primes; while the common point of the primes is quadruple for V^6. We shall consider this primal in the next section, showing that it is unirational but not birational; FANO has incorrectly stated that it is not even unirational.

9. Questions of irrationality. We have seen in § 7 that the threefolds W^{2p-2} of the first species are birational for $p = 7, 9, 10$ and unirational for $p = 4, 5, 6, 8$. The question now arises as to whether, in these latter cases, W^{2p-2} is irrational (non-birational), and also whether W^4 ($p = 3$) is not merely not birational but not unirational as well.

As early as 1908 FANO became convinced that both W^4 and W^6 are irrational (at that time it was not yet known that W^6 is unirational); in his first work (FANO [5]) on the subject he proposes to demonstrate the fact by proving that neither threefold contains a homaloidal system of surfaces. Such a system must be cut out by primals of a certain order, and will possess multiple base elements (either points or curves or both); and FANO asserts without proof that the relations he obtains, which certainly hold when these base elements are all distinct, will continue to hold when some of them are proximate. A second criticism concerns the lowering of the arithmetic genus of a surface by a multiple point or curve: as the theory of space curves suggests, the effect produced by such a multiplicity on a surface in higher space may well be different from that expected by analogy with the theory of surfaces in S_3.

In his second work (FANO [7]) FANO's object is to show that neither W^4 nor W^6 can contain a linear system of regular surfaces of genera and plurigenera unity more ample than that of the prime sections; since, in S_3, there exist systems of such surfaces having freedom 34 at least, it would follow that W^4 and W^6 are not only irrational but birationally distinct. FANO begins by remarking that a system of the required type may possess an adjoint surface (possibly of order zero) but that all the successive adjoints must be lacking; also that, if the system is cut out by primals of order n (>1), there must be a multiple base point of order greater than $2n$ or a multiple base curve of order greater than n, since otherwise no condition would be imposed on the n-th adjoint surfaces. Here the previous criticism concerning the evaluation of the arithmetic genus of a surface in higher space still applies. With regard to the base elements, FANO makes a number of unproved assertions about the effect of infinitesimal base points and curves on the successive adjoint surfaces; at that time no detailed study of proximate points, even of plane curves, had yet been published: subsequent results indicate that the question is even more complex than might have been imagined (cf. B. SEGRE [11]).

If FANO's statements could be justified, then it would follow that, on W^4, the system of prime sections is the sole system of the requisite type of freedom 2 at least, so that W^4 can admit no birational self-transformations other than collineations. On W^6, however, such transformations exist; thus the planes through any one of the lines of W^6 meet the threefold again in point-pairs of an involution I_2 which provides such a transformation. In this case FANO's result would be that, on W^6, the only systems of the required type, of freedom 2 at least, are the system of prime sections and its transforms by the process just indicated.

In 1912 ENRIQUES [6] showed that W^6 is unirational; and, on the basis of FANO's work, deduced from this particular example the existence of irrational involutions in S_3. This for long remained the standard (indeed the sole) example of such involutions.

In 1942 FANO [17] concluded an examination of the remaining types ($p = 5, 6, 8$) of the first species: in this work it is proposed to show that the threefolds in question cannot contain systems of surfaces of genera unity by means of which they can be mapped on the prime sections of a birational threefold, e. g. the W^{12} of the first species. It would seem that FANO had by then recognised the inadequacy of his treatment of the case of infinitesimal base elements, for he begins by assuming explicitly that, in the correspondences between his threefolds and W^{12}, there are no proximate base elements whatever: however, he gives no warning concerning the consequent lack of generality in his results, and concludes that the threefolds W^{2p-2} of the first species ($p = 5, 6, 8$) are all irrational. From this result he deduces that the general cubic primal of S_4 is likewise irrational, for he had previously shown (FANO [9]) that it is birationally equivalent to the W^{14} of the first species*). It had of course long been known that the cubic primal is representable on an I_2; in SNYDER [2] the algebra of the representation is given in detail. An account of the literature bearing on the problem of the cubic primal will be found in SNYDER (a) (see also VI, 8).

A more difficult question than the irrationality is that of the non-unirationality of the general quartic primal of S_4. This primal is certainly unirational, and representable on an I_6, as soon as it is specialised to contain a plane; and, as we have elsewhere remarked (IV, 5), it is representable on an I_4 if it is made to contain a cubic scroll. Although no general criterion of birationality of the quartic primal is known, many birational types have been discovered by various means. In the first place, the quartic primals with double plane, quadric, cubic scroll or tacnodal line are all birational, by FANO's theorem (IV, 11). Next we have the quartic primal containing three skew planes, first studied by D'AMICO [1] and then by ROTH [10]. More generally, it is obvious that a quartic primal is birational if it contains a surface endowed with a single apparent triple point; the principal types of such a primal have been obtained by TODD [14]. The projectively generated, or determinantal, quartic primal, with 20 nodes, was first considered, in a slightly special form, by ROTH [8]; the case where the determinant is symmetric is dealt with in ROTH [9]. Other particular determinantal types, including those with 45 nodes**), have been studied by BABBAGE [1] and TODD [11–13]. The general types which can be mapped on S_3 by systems of quartic surfaces have been investigated by BABBAGE and TODD [1].

Another question of a similar type which still awaits an answer concerns the unirationality, or otherwise, of the double S_3 with a non-singular sextic branch surface; using the methods of the previous chapter it is easy to construct unirational types for which, however, the branch surface has various singularities; another type, the unirationality of which is by no means obvious *a priori*, is obtained by projecting a W^6 of the first species from a line of itself: this gives a double S_3 with a sextic branch surface endowed with 31 nodes, images

*) By means of its quadric sections this primal may also be mapped on a W^{24} of the second species.

**) In this connection see BAKER (b).

of the lines of W^6 which meet the vertex of projection, and with other particularities whose influence on the result is not easily explained.

The presumed non-unirationality of the general V_3^4 of S_4 may be set in relation with the result (MORIN [8]) that the general quartic primal V_d^4 is unirational for all $d \geqq 5$. It may well be that the double S_d with general sextic branch variety V_{d-1}^6 is unirational for all sufficiently large values of d; but no light has yet been thrown on this question.

We turn now to the problem of constructing completely regular threefolds, on which adjunction terminates, which are unirational but not birational – or, what is the same thing, of establishing the existence of irrational involutions in S_3. We have already seen (in § 2) how to construct completely regular threefolds, on which adjunction terminates, which are neither birational nor unirational, by appealing to the theory of divisors; we shall now see that this theory serves equally well in the present circumstances and, moreover, that it enables us at the same time to construct irrational involutions in a space S_r of any dimension $r \geqq 3$.

For this purpose we consider the ENRIQUES primal V^6 introduced in § 8; as already mentioned, V^6 has six double planes, intersections in pairs of four primes in general position, four triple lines and a quadruple point O which is the point of concurrence of the primes in question. If these primes have equations $x_i = 0$ $(i = 1, 2, 3, 4)$, the equation of V^6, in homogeneous coordinates $(x_1, x_2, x_3, x_4, x_5)$, may be written in the form

$$\begin{aligned} x_1 x_2 x_3 x_4 \{x_5^2 + x_5 f_1(x_1, x_2, x_3, x_4) + f_2(x_1, x_2, x_3, x_4)\} \\ + g_2(x_2 x_3 x_4, x_3 x_1 x_4, x_1 x_2 x_4, x_1 x_2 x_3) = 0 \,, \end{aligned} \tag{1}$$

where f_i and g_i denote forms of order i in their respective arguments.

In the first place we observe that V^6 has divisor $\sigma = 2$. For consider the pencil of quadric cones represented by the equation $x_1 x_2 = \lambda x_3 x_4$, which pass through the double planes of V^6 incident in lines to a pair C_1, C_2, say, of skew double planes; this pencil cuts on V^6, residually to the four double planes in question, a linear pencil $|C|$ of DEL PEZZO quartic surfaces. Evidently we have $2C_1 \equiv 2C_2 \equiv C$; moreover, the surfaces C_1, C_2 must be isolated, since otherwise they would belong to a pencil of quadrics, which cut the generic prime section – an ENRIQUES surface – in conics, so that the latter would be rational or scrollar. It follows that $\sigma \geqq 2$; and since $\sigma \leqq \sigma_0$, where σ_0 is the divisor of a prime section of V^6, we must have $\sigma = 2$.

Next, we show that V^6 is unirational and representable on an involution I_4. In fact, through any simple point of V^6 there passes a unique plane meeting C_1 and C_2 in lines, and therefore cutting V^6 residually in a conic, in general irreducible. We thus have on V^6 a rational congruence of conics with index unity; whence, by § 2, V^6 is completely regular, with genera and plurigenera zero. Since these conics are quadrisecant to any fixed irreducible surface of the pencil $|C|$, it follows from ENRIQUES's theorem (IV, 5) that V^6 is unirational and representable on an I_4; and since, as we have seen, V^6 has divisor greater than unity, it cannot be birational.

It follows that *there exist irrational space involutions of order four.* More generally, we may deduce from this result that *there exist irrational involutions of order four in any space S_r $(r > 3)$.* For consider the cone which projects V^6 from any given vertex S_{r-4} $(r \geqq 4)$; evidently this is unirational and representable on an I_4. Also, by the above result, and the LEFSCHETZ formula (III, 4) for the divisor of a product variety, the cone has divisor 2, and is therefore irrational*).

It is interesting to note that the ENRIQUES sextic surface has thus served the same useful purposes in the theory of threefolds as it had previously done in the theory of surfaces. It has already appeared in IV, 2 in connection with the construction of non-unirational threefolds: we now give two further illustrations of its applications to the present theory.

First, we observe that, if the coefficients in (1) are allowed to vary, we obtain a linear system of ENRIQUES primals with double planes in common. Now among these primals are cones which project ENRIQUES surface from O; these are obtained from (1) by suppressing the terms in x_5. Such cones, which are in effect products of a rational curve and an ENRIQUES surface, are completely regular, with genera and plurigenera zero (as follows from the formulae of I, 11); and, by IV, 4, they are not unirational. We have thus an example of *a linear system of threefolds, the generic member of which is unirational, containing particular threefolds which are completely regular, with genera and plurigenera zero, but which are not unirational.*

In the second place, if we project V^6 from O on to the prime $x_5 = 0$, we have a representation on a double space S_3 of which the branch surface is given by the equation

$$x_1 x_2 x_3 x_4 \{x_1 x_2 x_3 x_4 (f_1^2 - 4 f_2) - 4 g_2\} = 0 . \tag{2}$$

This consists of an ENRIQUES surface together with the four planes of the associated tetrahedron; and furnishes an illustration of *a double S_3 which is unirational but not birational.*

The above results are given in ROTH [25], where a second example of such a double space is constructed; this contains a pencil of rational surfaces of the second family, with rational bisecant curve, and is therefore unirational. It also contains a web of ENRIQUES surfaces of which there are two disequivalent submultiples, so that the threefold mapped on this double space has divisor greater than unity, and thus is not birational.

*) Most of the above results can actually be obtained without using the theory of the base: see ROTH [38].

10. Some unsolved problems. In this chapter and also in the preceding we have mentioned incidentally some of the more important questions which still await a solution; in the present section we review these and other problems systematically.

I. In order to construct completely regular threefolds of plurigenera zero which are not unirational we have had recourse to the theory of the divisor: it remains to construct examples which have unit divisor – one of these is presumably the general quartic primal of S_4.

II. Presumably there exist infinite series of unirational threefolds which are birationally distinct; one such series might be found by constructing unirational threefolds with different divisors. But, as before, there will then remain the problem of finding unirational types with unit divisor, among which (we may suppose) will be certain of the FANO threefolds; we require here to discover some new invariant which would serve to distinguish the unirational from the birational types.

III. It is known that, on any surface of plurigenera zero, the adjunction process always terminates; however, no direct proof of this result has yet been given – the customary demonstration involves the construction of a pencil of rational curves on the surface. Similarly we may ask whether, on a threefold of plurigenera zero, the adjunction process always terminates; in this case a quite different procedure is called for. An associated problem is the reduction to birationally distinct standard forms of the threefolds on which adjunction terminates (such threefolds include of course the unirational types): what are the characters appropriate to such a classification?

IV. We come now to problems concerning threefolds generated by rational congruences of rational curves. First there is the problem of constructing such a congruence, with index unity, which does not admit a rational plurisecant surface; as previously suggested, an example might be provided by the V_3^n of S_4 whose only singularity is an ordinary $(n-2)$-ple line $(n > 4)$.

Next, we have seen that, if a threefold contains a rational congruence, of index unity, of rational curves, it is completely regular: does this conclusion hold in the case where the congruence has index greater than unity?

This last question has a bearing on the theory of unirational threefolds. We have seen that such threefolds have plurigenera zero and are superficially regular: it has still to be shown by classical methods that they are also tridimensionally regular. This might be achieved by proving that the adjoints to any given surface on a unirational threefold cut a complete system on the surface. But if the previous question could be answered in the affirmative, the required result could then be inferred; in fact, from the parametric representation of the threefold, which is of the form

$$x_i = f_i(u, v, w) \quad (i = 1, 2, 3, \ldots), \tag{1}$$

where the f_i are rational functions of the parameters u, v, w, we see that corresponding to values v = constant, w = constant, we have a rational congruence of rational curves; in general, however, this has index greater than unity.

We may now ask: does every unirational threefold contain a rational congruence, of index unity, of rational curves? The analogous result in the theory of surfaces constitues the proof of CASTELNUOVO's theorem concerning the rationality of plane involutions; and, by § 2, if the question could be answered affirmatively, it would follow that every unirational threefold is completely regular. It can be shown (ROTH [32]) that, in the particular case where, in (1), the surfaces of the system u = constant, and also those of the system v = constant, are generally non-singular, and where the intersection of the two systems is simple and non-singular, the existence of a congruence of index unity can be inferred. A similar result holds for unirational varieties of any dimension; however, such hypotheses may well be restrictive.

V. We know that the invariants of any surface which contains an irrational pencil of rational curves are at once determinable when the genus of the pencil is given. In the case where a threefold contains a congruence, of index unity, of rational curves, the genera, plurigenera and superficial irregularity are calculable in terms of the invariants of the congruence, provided the latter possesses a unisecant surface (I, 11). Supposing, however, that such a unisecant does not exist, it should still be possible to determine formulae for the invariants, or at least limits between which they must lie.

VI. It seems likely that a threefold may contain a rational pencil, or even a net, of rational surfaces without being unirational. A threefold which contains a rational pencil of rational surfaces certainly has plurigenera zero: the question arises whether it must be completely regular. Now when the surfaces are of the first family the threefold is birational; and when they are of the third family we know that there is a rational congruence, of index unity, of rational curves, so that in these cases the result is immediate. In all the remaining cases the threefold is either birational or is representable on a double space with a branch surface of one of the standard types; and if this surface has no other singularities than those prescribed in the "general" case, we find that the double space has arithmetic genus zero and is superficially regular. But there is no reason why the branch surface should not acquire further singularities; and to deal with this case we should need a knowledge of the theory of proximate points which at present we do not possess.

VII. We have already mentioned the problem of the cubic primal in S_4; this will reappear in the next chapter. Assuming that the primal is in general irrational we can, however, assert that it becomes birational upon acquiring a node. Consider now the cubic primal in S_5; in general this contains no planes; if it acquires two skew planes, it can be projected birationally on to S_4; and the interesting fact is that such a primal is in general non-nodal. Again, if the primal contains a rational normal quartic scroll it is certainly birational, and for a similar reason (since the scroll possesses one apparent double point). By a simple counting of constants we should be led to infer that a cubic primal of S_5 always contains ∞^1 such scrolls: however, FANO [16] has shown that this is not the case. FANO proves that, if the cubic primal contains one rational normal quartic scroll, it must contain an ∞^2 system of such scrolls, from which one infers that in general it contains none. The problem is intimately related to the study of the FANO threefolds, since any quadric primal passing through such a scroll meets the cubic primal in a particular type of W^6.

Chapter VI.

Continuous Transformation Groups.

1. Groups of auto-collineations. The present chapter is mainly concerned with varieties, and in particular threefolds, which admit finite continuous groups of birational self-transformations or automorphisms. The general theory of such groups is due to LIE*); the results of its application to curves and surfaces, which are classical, underlie all that follows, and will be stated as and when they are required.

Let G be an algebraic continuous group of automorphisms of a given variety V_d ($d \geqq 1$). Such a group may or may not be *permutable* (commutative, Abelian); again, it may be either *transitive* – completely or generally**) – or *intransitive*; and it may be *integrable* or *non-integrable.* We recall that an ∞^r continuous group G_r is said to be integrable if it contains a set of subgroups G_i ($i = r-1, r-2, \ldots, 1$) each dependent on a number of essential parameters equal to the index i, and such that each G_i is an invariant subgroup of G_{i+1}. The group G_r is non-integrable if it does not contain a set of subgroups G_i with the properties just described. Such a group must contain at least three essential parameters; and every non-integrable continuous group is *simple,* i. e. contains no invariant subgroup. It may further be shown that a continuous group is non-integrable if and only if it contains at least an ∞^3 simple group.

*) A useful account of this theory can be found in ENRIQUES-FANO [1].

**) In the sequel "transitive" will mean "generally transitive".

It is important to note the geometrical significance of these results in the case where G_r is a group of collineations of S_r; if G_r is integrable, then it possesses at least one fixed united point, one fixed united line through this point, . . . , and, finally, one fixed united prime containing all these subspaces. The converse of this result is true.

The non-integrable collineation groups of S_r have been studied by FANO [1], with special attention to the ∞^3 simple groups. FANO obtains the equations of such a group, and establishes the fundamental result that *every* ∞^1 *subgroup contained in such a group is algebraic.*

We begin with the cases where G is a group of auto-collineations, or projective transformations, and where $d = 1, 2$. First, it may be proved (SEVERI, d) that any curve of S_r which admits a continuous group of ∞^2 auto-collineations is a rational normal curve of S_r; and that any *algebraic* curve which possesses a continuous ∞^1 collineation group is rational. Next, we have the theorems (ENRIQUES, a): I. any surface which admits a continuous group of auto-collineations is rational or scrollar; II. any surface endowed with a continuous transitive (hence at least ∞^2) group of auto-collineations is rational.

Using these results, FANO [1, 2] classified the threefolds which admit either a continuous integrable ∞^4 group of auto-collineations or a continuous non-integrable group of auto-collineations. It appears from FANO's analysis that such threefolds form a very restricted class. Shortly afterwards FANO [3] proved the theorem:

Any threefold which possesses a transitive (hence at least ∞^3*) continuous group of auto-collineations is birational.*

Let V be the threefold in question and suppose, in the first place, that the group G is integrable; then G contains at least one invariant ∞^1 subgroup which, by the general theory (§ 6) can be assumed to be algebraic, and which therefore possesses rational trajectories $\mathfrak{C}$, say. These form an algebraic congruence Γ (of index unity*)) which certainly admits unisecant surfaces; for it is known that the united points which this subgroup defines on the generic $\mathfrak{C}$, if distinct, will describe two surfaces, each in consequence unisecant to Γ; while, if these united points are coincident, there is nothing to prove. If, then, we can show that Γ is rational, it will follow that V is birational.

Now by applying the transformations of G to this unisecant surface we can construct an infinity of simple linear systems of unisecants of arbitrarily high dimensions. By means of any one such system, which is invariant under G, we can map V birationally on a threefold V' on which there is a congruence Γ' of lines $\mathfrak{C}'$, corresponding to the curves $\mathfrak{C}$; and the group G thereby corresponds to a group G' which operates projectively and transitively on Γ'. Hence Γ' can be regarded

*) Throughout this chapter all congruences are of index unity.

as a surface (the points of which are mapped by curves $\mathfrak{C}'$) endowed with a transitive collineation group; whence, by the above theorem II, Γ' is rational and so Γ is likewise rational.

Suppose, in the second place, that G is non-integrable; then it contains an ∞^3 simple subgroup, in which every ∞^1 subgroup is algebraic. Hence G contains algebraic subgroups whose trajectories are rational curves $\mathfrak{C}$; for each such subgroup we have a congruence Γ of curves $\mathfrak{C}$ which, as before, possesses unisecant surfaces. If all these congruences were coincident, the unique congruence so obtained would be invariant under G and could be dealt with as before. Supposing that this is not the case, let Γ, Γ' be two of these congruences, generated respectively by curves $\mathfrak{C}$, $\mathfrak{C}'$; and consider the curves $\mathfrak{C}$ which meet a given curve $\mathfrak{C}_0'$. These curves form a rational ∞^1 system, since they cut $\mathfrak{C}_0'$ in sets of points which obviously form an involution. We can thus construct an infinity of surfaces A, each of which contains a rational ∞^1 system of curves $\mathfrak{C}$; there are now two possibilities:

1. If the curves $\mathfrak{C}$ which meet $\mathfrak{C}_0'$ meet a finite number of curves $\mathfrak{C}'$, the surfaces A form a doubly infinite system.

2. If the curves $\mathfrak{C}$ which meet $\mathfrak{C}_0'$ meet an infinity of curves $\mathfrak{C}'$, the surfaces A form a simply infinite system.

In case 1 there are ∞^1 surfaces A through any given $\mathfrak{C}$; these constitute a rational system Σ, for each such surface either contains a single curve $\mathfrak{C}'$ incident to $\mathfrak{C}$ or a set of an involution of these curves which, by what has been said, form a rational ∞^1 system. Hence, regarding Γ as a surface S, and the ∞^1 system of curves $\mathfrak{C}$ on a given surface A as a curve on this surface, we have the result that S contains an ∞^1 rational system of rational curves; thus, by a theorem of Castelnuovo (IV, 6), S is rational. Thus Γ is rational and so V is birational.

In case 2 we observe that the ∞^1 surfaces A form a pencil; for the surface A which passes through the general point of V is uniquely specified since it contains the unique curve $\mathfrak{C}$ through this point, and hence all the curves $\mathfrak{C}'$ incident to $\mathfrak{C}$. If now there exists on V a congruence Γ'' analogous to Γ and Γ' whose curves $\mathfrak{C}''$ do not lie on A, the pencil $\{A\}$ must meet each curve $\mathfrak{C}''$ (which is of course rational) in sets of an involution; hence the pencil is rational, and hence Γ is likewise rational. If instead $\{A\}$ belongs to every congruence such as Γ, Γ', it is necessarily invariant under G; and since G is transitive for V, the pencil must be rational. This last remark is justified as follows. Any continuous system Σ of varieties A of a space S_r can be regarded as a variety M, say, in some space, such that the auto-collineations of Σ are mapped by auto-collineations of M. For each variety A can be taken to be the complete intersection (or base variety) of a linear system of primals of any sufficiently high order, so that we can replace the varieties A by linear systems of such primals. These in their turn can be regarded as subspaces; and in such a mapping the projective character of the collineations in S_r will be preserved.

Consider, then, this representation applied to the pencil $\{A\}$; we thus see that $\{A\}$ can be regarded as an algebraic curve endowed with at least ∞^1 auto-collineations. Hence the curve is rational and so, once more, Γ is rational.

2. Automorphisms in general. We know from particular examples that the problem of characterising those varieties V_d which are invariant under a given finite continuous group G of automorphisms cannot in general be reduced to the preceding; however, in the case where V_d is *superficially regular*, such a reduction is always possible. For consider any sufficiently ample simple linear system of hypersurfaces on V_d; applying to it all the transformations of G we obtain a *linear* system (I,6) of hypersurfaces which is simple and also invariant under G; and V_d may thereby be mapped on a variety V'_d which admits a group G' of auto-collineations.

This result applies in particular to the rational curves; in fact it can be proved that any curve which admits a continuous group of ∞^2 automorphisms is rational. But the case of ∞^1 automorphisms is essentially different; for, apart from the rational curves, we have the elliptic curves*) which admit continuous groups of ∞^1 automorphisms. As regards curves, the theory is completed by the SCHWARZ-KLEIN theorem that an algebraic curve of genus greater than unity can admit only a finite number of automorphisms (SEVERI, d).

Turning to the case of the surfaces, we first recall the following results (ENRIQUES, a):

1. Any surface which admits a rational ∞^1 group of automorphisms is rational or scrollar.
2. Any surface which admits a continuous group of automorphisms with an invariant linear system of curves is rational or scrollar.
3. A surface F is rational or scrollar if it possesses a continuous group of automorphisms, of dimension $r \geqq 2$, which operates intransitively on F, or of dimension $r \geqq 3$, which operates transitively on F.
4. A surface is rational or scrollar if it admits a continuous *series* of automorphisms which do not generate a finite group.

The proof of 1. is immediate, for the trajectories of the group form a pencil of rational curves. With regard to 2., where it is understood that the linear system of curves is invariant as a whole, we first establish the existence of a pencil of trajectories, which *a priori* may be rational or elliptic; the latter possibility is excluded by the fact that

*) The two cases may often be distinguished in practice by the circumstance that, if the group leaves invariant a linear series of freedom $r \geqq 1$, the curve must be rational.

each trajectory contains an invariant linear series. The same considerations may be applied to 3. Theorem 4, which belongs to a different category of problem, is proved by showing that the surface in question must contain a curve system whose characters n, p satisfy the inequality $n > 2p - 2$.

We now pass to the problem of determining the non-scrollar surfaces which admit continuous groups of automorphisms. We suppose that the group G in question is algebraic and not contained in a more ample group; there are then two possibilities:

a) G operates intransitively on the points of the given surface; there is then a pencil of trajectories which, by the preceding results, must be elliptic curves.

b) G operates transitively on the points of the surface F, and thus has dimension 2. To begin with, we may assume that the transformations of G are all permutable, since otherwise the transformations which are permutable with a given transformation form an ∞^1 subgroup, and we are led to a special case of a). Next, we may suppose that G is *completely* transitive; for otherwise we should have a united point or curve on F, in which case F would be rational or scrollar. Finally, it is easily seen that G must be *simply* transitive, i. e. that there is a unique transformation of G which carries any given point of F into any other given point.

Supposing, first, that this is the case, we know that F is a PICARD surface, or hyperelliptic surface of rank 1; it is then a particular case of the PICARD varieties. The second possibility is ruled out by remarking that, if there are n (>1) transformations of G which carry a given point of F into another given point of F, we can set up a $(1, n)$ correspondence between the points of F and those of a second surface F' which must be of PICARD type; and since this correspondence is without branch elements, we deduce that F is a PICARD surface, and that the transformations of G are of the first kind (§ 3).

Leaving the PICARD surfaces for the present, we return to case a). A surface which admits an elliptic group of automorphisms is called *elliptic*; the trajectories of the group form a pencil, of some genus $\varrho \geqq 0$, of elliptic curves $\mathfrak{C}$; the irreducible members of this pencil are birationally equivalent, but the pencil will in general contain a certain number of multiple elliptic members. It may be proved that the surface also contains an elliptic pencil of birationally equivalent curves $\mathfrak{K}$, of genus $\pi \geqq 1$; and, conversely, that any surface containing a pencil $\{\mathfrak{C}\}$ is necessarily elliptic. In the classification of the elliptic surfaces the number $[\mathfrak{C}\mathfrak{K}] = d$, called the determinant, plays an important part. The first geometrical discussion of these surfaces was given by ENRIQUES (a); they had, however, been previously been considered by PAINLEVÉ (a) from the group-theoretic point of view.

While the PICARD surfaces possess a pure canonical curve of order zero, the elliptic surfaces have a canonical curve, effective or virtual, compounded of the pencil $\{\mathfrak{C}\}$. The types $\varrho = 0$, $\pi = 1$, which possess a virtual canonical curve of order zero, have a natural parametric representation by means of hyperelliptic functions; we therefore call them *improperly hyperelliptic* (or Abelian). It may be shown that the PICARD surfaces and the elliptic surfaces have absolute linear genus 1 and arithmetic genus -1; and that any surface with arithmetic genus -1 must belong to one or other of these two classes – where, for convenience, we include the elliptic scrollar surfaces, given by $\pi = 0$ (ENRIQUES, a). It may also be proved (DANTONI [1]) that these surfaces are characterised by the property of having SEVERI series of order zero.

The elliptic surfaces are the simplest examples of the pseudo-Abelian varieties (first discussed briefly by DANTONI [2]) which will be considered in § 4.

3. Abelian and quasi-Abelian varieties. The concepts of PICARD surface and elliptic surface readily generalise to varieties of any dimension, with highly interesting results. As regards the PICARD variety, there already exists a considerable literature (see ANDREOTTI, a; CONFORTO, a; LEFSCHETZ, a) to which we may refer for proofs of the results required here.

An *Abelian variety* W_p is an irreducible variety the coordinates of whose general point are expressible as Abelian functions, of genus p, of p independent variables $u_1, u_2, \ldots, u_p$. The *rank* r of W_p is defined as the number of points in the primitive period parallelepiped U_p which correspond to the general point of W_p. In the case $r = 1$, W_p is called a PICARD variety.

Assuming, as is always possible (SIEGEL; BARSOTTI [1], KODAIRA [5]) that the PICARD variety has been cleared of singularities, we can proceed to remove any exceptional subvarieties that it may contain; as a result of these two operations we obtain a model whose points are in (1, 1) unexceptional correspondence with those of U_p; we shall denote this throughout by V_p.

Every variety V_p possesses two types of automorphism, the so-called ordinary transformations; and if it is *general* (i. e. has general moduli) it admits only these. The transformations of the first kind are represented by the equations $u'_i = u_i + a_i$ $(i = 1, 2, \ldots, p)$, where the a_i are arbitrary constants; evidently these form a completely transitive permutable continuous ∞^p group. On the other hand, the transformations of the second kind, represented by the equations $u'_i = -u_i + a_i$ $(i = 1, 2, \ldots, p)$, do not form a group. More generally, we may consider series of sets of m points on V_p, given by equations of the form

$$u_i^{(1)} + u_i^{(2)} + \cdots + u_i^{(m)} = c_i \quad (i = 1, 2, \ldots, p)\,. \tag{1}$$

It may be shown (ZARISKI, a) that any series of linear circulation zero, and hence, in particular, any series of intersection on V_p, is representable by equations of this type.

A general variety V_p contains no PICARD varieties V_q $(q < p)$; it can be shown that, if V_p contains one such variety V_q, it must contain a congruence, free from base points and singular points, of varieties V_q, the congruence itself being of PICARD type; and, further, that V_p must then contain a second PICARD congruence of PICARD varieties V_{p-q}. We shall call such a variety V_p *special of type* q (or $p - q$).

A particularly notable subclass of PICARD varieties is constituted by the JACOBI varieties; a JACOBI variety J_p is the manifold which maps the sets of p points of a curve (in general irreducible) of genus p. It may be shown that J_p can be general as a PICARD variety only for $p \leqq 3$.

We consider now some important characteristic properties of V_p. We have remarked that V_p possesses a completely transitive permutable continuous group of ∞^p automorphisms; conversely, it may be proved that any variety W_p which admits such a group must be a PICARD variety of genus p. It is obvious from the parametric representation that V_p has superficial irregularity p; hence we may add that any variety W_p which admits the group in question must have superficial irregularity p.

ENRIQUES (a) has shown that *any variety which can be mapped on a multiple* V_p *without branch points is itself a* PICARD *variety*; the proof consists in showing that the variety possesses the above-mentioned group of automorphisms.

From this result we may deduce (ROTH [33]) that any variety W_p of superficial irregularity p, with a pure canonical hypersurface of order zero, and not containing a congruence of irregularity p, is a PICARD variety. The method of proof, already employed in III, 5, is to map a continuous system of hypersurfaces of W_p by the points of a PICARD variety.

We have already alluded to the fact that V_p has a canonical hypersurface of order zero; this property is included in the general result, due to SEVERI, that *each of the canonical varieties* $X_k(V_p)$ $(k = 0, 1, \ldots, p-1)$ *has order zero.* (SEVERI has conjectured that such a property characterises the PICARD varieties, but this has been verified only for $p = 2$.) It follows from this that all the canonical invariants (II, 5) of V_p are zero.

Finally, we observe that LEFSCHETZ [2], using the postulational definition, has shown that the arithmetic genus of V_p is equal to $(-1)^{p-1}$.

Turning now to the Abelian variety W_p of rank $r > 1$, we see from the definition that W_p may be mapped by an involution of order r on a PICARD variety V_p. In general V_p will have divisors greater than unity; in that case, denoting the product of the divisors by s, we can proceed to map W_p by an involution I_n, of order $n = rs$, on a PICARD variety with unit divisors, which we may continue to denote by V_p. Since W_p is irreducible, I_n is necessarily simple.

The classification of the varieties W_p thus turns on that of the involutions I_n. To begin with, we have the following results (ROTH [34]):

I. *The pure canonical and pluricanonical hypersurfaces of* W_p, *if effective, are all of order zero.*

II. If I_n *possesses* ∞^{p-1} *coincidences, the geometric genus and plurigenera of* W_p *are all zero.*

The group-theoretic method of classifying the varieties W_p (which is the only means at present known) is classical for $p = 2$; it depends on the theorem (ANDREOTTI, [2]): *If* W_p *has some plurigenus greater than zero, then* I_n *can be generated by a finite group* G_n *of automorphisms of* V_p. Thus the only general cases which can be treated by this method are those in which I_n has at most ∞^{p-2} coincidences.

It may now be shown that G_n can be generated by a finite set of substitutions of the form

$$u'_i = \sum_{j=1}^{p} a_{ij} u_j + b_i \quad (i = 1, 2, \ldots, p), \tag{2}$$

where a_{ij} and b_i are constants. In the case where W_p has superficial irregularity $q > 0$, we may show further that q of the above relations may be taken to be

$$u'_i = u_i + b_i \quad (i = 1, 2, \ldots, q). \tag{3}$$

LEFSCHETZ [2] has remarked that, by modifying suitably the period matrix of V_p, the remaining relations of the set (2) may be reduced to the form

$$u'_j = \varepsilon_j u_j + b_j \quad (j = q+1, q+2, \ldots, p). \tag{4}$$

The constants ε_j are called the multipliers of the substitution; since the group generated by (3) and (4) is finite, these must be roots of unity, other than unity itself.

Equations (3) suggest the result: *every Abelian variety* W_p *of superficial irregularity* p *is a* PICARD *variety*. This has been proved by SEVERI [9].

A variety W_p is called *properly* or *improperly* Abelian according as it does not or does admit a parametric representation by means of Abelian functions of genus (or genera) lower than p. Among the proper types, other than the PICARD variety itself, we have the WIRTINGER variety, which is superficially regular, and which maps the involution I_2 generated by the substitution $u'_i = -u_i + a_i$, on V_p.

Leaving further developments of this topic for the present, we consider next how the notion of PICARD variety may be generalised. Evidently this may be effected in two different ways: in the first place, we may envisage a variety V_π which is endowed with a permutable continuous ∞^π group of automorphisms, where now the group is only *generally* transitive; we thus arrive at the concept of *quasi-Abelian variety*, introduced by SEVERI (b), which will arise naturally in a later connection. In the second place, we may suppose that the group has some dimension q less than that of the variety itself; such a variety, which is a generalisation of the elliptic surface, we call *pseudo-Abelian of type q* (ROTH [33]). Actually, we shall see that V_π is merely a special case of the pseudo-Abelian variety.

4. Pseudo-Abelian varieties. Consider a non-singular variety W_p which admits a permutable continuous group G of ∞^q automorphisms $(1 \leq q \leq p-1)$. The trajectories of G constitute a congruence $\{V_q\}$ of varieties V_q, the generic member of which is irreducible; and all irreducible members are birationally equivalent to G and hence to each other*). Each such variety V_q is invariant under G and no two trajectories intersect. If we assume that G acts transitively, without exceptions, on the generic V_q, it follows from § 3 that V_q must be a PICARD variety; if, further, we assume that V_q represents biunivocally and unexceptionally the transformations of G, then V_q is non-singular and free from exceptional subvarieties. For reasons of simplicity, which will appear later, we shall suppose that V_q is general; this restriction is removed in ROTH [36]. But in any case the operations of G must be transformations of the first kind on V_q itself (§ 2).

We first prove that W_p *contains a* PICARD *congruence* $\{V_{p-q}\}$ *of* ∞^q *birationally equivalent varieties* V_{p-q}. For, consider the series of intersection cut on V_q by the primals of any given order in the ambient space of W_p; to any set of this series, which is represented by equations of type (1), there corresponds a unique set of the analogous series cut on any other irreducible trajectory V'_q, and vice-versa. Now by (1), there is a finite number n $(=2m^{2q}$, where m is the order of the series) of transformations of the first kind which leave invariant each such series and which transform V_q birationally into V_q; to each point P of V_q we may thus make correspond a unique set of n points of V'_q, and to each point P' of V'_q there corresponds a unique set of n points of V such that P (or P') belongs to a unique set of such points. As V'_q varies in $\{V_q\}$, the set describes a variety of dimension $p-q$, possibly reducible; if it is irreducible, denote it by V_{p-q}, while if it is reducible, let V_{p-q} be any one of its components. Then the transforms of V_{p-q} under G form an irreducible system of index 1, which is the required congruence $\{V_{p-q}\}$; its members cut V_q in sets of an involution which is without coincidences and which, since V_q is general**), must be of PICARD type. Hence $\{V_{p-q}\}$ is a PICARD congruence, which is evidently free from base points and

*) This result is established exactly as in the theory of the elliptic surfaces.

**) If V_q is not general, this need not be the case (see § 5).

singular points. We shall assume that the generic V_{p-q} is non-singular, from which, as we shall see, we are led to suppose that every V_{p-q} is non-singular.

The intersection number $[V_q V_{p-q}] = d$, called the *determinant* of W_p, is an important character of the variety. It should be noted that each variety V_{p-q} is invariant under a permutable group G_d of automorphisms, namely the transformations of G which leave each V_{p-q} invariant.

Conversely, *any variety W_p which contains a congruence $\{V_q\}$ as specified below is pseudo-Abelian of type q*; the proof is exactly as in the theory of the elliptic surfaces.

We next show that *W_p can be mapped on a d-ple pseudo-Abelian variety of determinant unity*. Leaving aside the case $d=1$, for which the result is obvious, we proceed thus. We first construct the variety $W_p^* = V_q^* \times V_{p-q}^*$, where V_q^* and V_{p-q}^* are birationally equivalent, without exceptions, to $\{V_{p-q}\}$ and $\{V_q\}$ respectively; such a variety contains two congruences which, without risk of confusion, we may denote by $\{V_q^*\}$ and $\{V_{p-q}^*\}$, where V_q^* is a trajectory of W_p^*. We now make correspond to the general point of W_p^* the set of d points common to the generic V_q and V_{p-q}, thus obtaining a mapping on the d-ple variety W_p^*. In this mapping each V_q corresponds to a d-ple PICARD variety V_q^*, the representation being without branch points, since the involution i_d cut by $\{V_{p-q}\}$ on the generic V_q is free from coincidences. Hence the branch locus is either lacking altogether or else consists of a number of irreducible varieties belonging to the congruence $\{V_q^*\}$, i. e. generated by varieties V_q^*; such varieties, which we shall suppose to be non-singular, may have any dimension from q to $p-1$, those of dimension q consisting of isolated V_q^*'s. To each generator $V_{q,s}^*$, say, of an $(s-1)$-fold component of the branch locus, there corresponds a variety $V_{q,s}$ which is an $(s-1)$-fold element of the coincidence locus of the involution I_d defined by the intersections of $\{V_q\}$ and $\{V_{p-q}\}$, and which is such that $sV_{q,s} \equiv V_q$; the numbers s $(2 \leqq s \leqq d)$ may *a priori* be any divisors of d.

Since $V_{q,s}$ is mapped on a d/s-ple PICARD variety $V_{q,s}^*$ without the intervention of branch points, it follows that $V_{q,s}$ must also be a PICARD variety (§ 3). As a hypothesis of generality we may assume that each variety $V_{q,s}$ is non-singular; in that case the representation of W_p on W_p^* has no other exceptional features than those already described; in particular, no trajectories other than the varieties $sV_{q,s}$ are reducible.

We now show that *the superficial irregularity q_2 of W_p satisfies the inequality**) $q_2 \geqq q_2^* + q$, where q_2^* denotes the irregularity of the congruence $\{V_q\}$. For the superficial irregularity of W_p^* is evidently the sum of the superficial irregularities of V_q^* and V_{p-q}^*. And (SEVERI [9])

*) DANTONI [2] shows by transcendental methods that $q_2 = q_2^* + q$.

the irregularity of the generic surface on W_p is at least equal to that of the generic surface on W_p^*.

Again, from this mapping, we obtain *an equivalence for the virtual canonical system* $|X_{p-1}|$ of W_p. In the first place, TODD [1] has shown, that, on W_p^*, $X_{p-1}^* \equiv X_{p-q-1}(V_{p-q}^*) \times V_q^* + V_{p-q}^* \times X_{q-1}(V_q^*)$. Since $X_{q-1}(V_q^*)$ has order zero (§ 3), $X_{p-1}^* \equiv X_{p-q-1}(V_{p-q}^*) \times V_q^*$. Hence, denoting by $\Sigma(s-1)W_{p-1}^*$ the composite branch hypersurface on W_p^*, we have, by the transformation law for canonical systems (ENRIQUES, a),

$$X_{p-1} \equiv \overline{X}_{p-1}^* + \Sigma(s-1)\overline{W}_{p-1}^* , \tag{1}$$

where the bar over a symbol denotes the transform of the variety in question. Thus $|X_{p-1}|$ contains as $(s-1)$-ple component every hypersurface of the coincidence locus of I_d which is generated by $V_{q,s}$'s, and passes $(s-1)$-ply through each variety of lower dimension which is generated by $V_{q,s}$'s. Hence, in all cases, $|X_{p-1}|$ belongs to the congruence $\{V_q\}$.

We may add that, in ROTH [36], equivalences analogous to (1) are given for the varieties X_k $(k \geqq q)$.

We have noted the fact that, for a PICARD variety, the canonical varieties of every dimension are of order zero; we shall now show that *the canonical systems* $\{X_k(W_p)\}$ $(k = 0, 1, \ldots, p-1)$ *of* W_p *either belong to the congruence* $\{V_q\}$ *of trajectories or else have order zero.*

This property is familiar in the case $p = 2$, of the elliptic surfaces; it will be established on the inductive hypothesis that it is true for every pseudo-Abelian variety of dimension less than p. For this purpose it is convenient to restate the definitions of the varieties $X_k(W_p)$ introduced in II, 5 in a more suitable form.

First, let $k = 0$; and let S be a hypersurface of a rational pencil $|S|$, of general character, on W_p, with Jacobian set δ; then X_0 is defined inductively, for $p = 2, 3, \ldots$, by the equivalence

$$X_0(W_p) \equiv \delta - 2X_0(S) - X_0(S^2) . \tag{2}$$

Precisely as in the theory of surfaces (ENRIQUES, a), we may now show that if instead S belongs to an irrational pencil $\{S\}$, of genus $\varrho > 0$, and of general character, then

$$X_0(W_p) \equiv \delta + 2(\varrho - 1) X_0(S) . \tag{3}$$

In the case $k > 0$, the varieties $X_k(W_p)$ are defined, by induction on k and p, by the equivalence

$$X_k(W_p) \equiv A_k(S) - X_k(S) \quad (k = 1, 2, \ldots, p - 1) , \tag{4}$$

where S is any hypersurface of W_p, and $A_k(S)$ is an *adjoint* variety to S, cutting S in a $X_{k-1}(S)$; and where $X_{p-1}(S) = S$.

The proof of our theorem depends on a lemma which generalises a well known result (ENRIQUES, a) in the theory of surfaces and which is established in exactly the same way:

I. Let $|S|$ be an irreducible pencil of W_p, with irreducible simple base variety S^2, and let S_0 be a member of $|S|$ whose only singularity is an ordinary h-ple irreducible non-singular variety W_r $(1 \leqq r \leqq p-2)$ which does not meet S^2; then the contribution made by S_0 to the number $[X_0(W_p)]$, calculated from $|S|$, is $(h-1)$ $[X_0(W_r)]$.

II. Let $\{S\}$ be a pencil, rational or irrational, of W_p, which is free from base points and which contains a member S_0 consisting of an irreducible non-singular variety W_{p-1} counted h times; then the contribution made by S_0 to the number $[X_0(W_p)]$ is $(h-1)$ $[X_0(W_{p-1})]$.

For the second part of the lemma we of course use (3). In either case, the method of proof shows that, if S_0 contains a finite number of such varieties W_r, no two of which intersect, the total contribution made by S_0 to $[X_0(W_p)]$ is the sum of the separate contributions.

Consider first the case $q = p-1$; here we have a pencil $\{V_{p-1}\}$, without base points, of PICARD varieties on W_p, containing a certain number of s-fold PICARD varieties $V_{p-1,s}$. By § 3, such varieties make no contribution to the value of $[X_0(W_p)]$, calculated from $\{V_{p-1}\}$; and since $\{V_{p-1}\}$ contains no nodal members, we deduce that $[X_0(W_p)] = 0$. Again, by (1), $|X_{p-1}|$ is compounded of the pencil $\{V_{p-1}\}$; and there is nothing more to prove.

Next, suppose that $1 \leqq q \leqq p-2$; and take $|S|$ to be a pencil of hypersurfaces belonging to the congruence of trajectories and endowed with an irreducible base variety S^2. Such a pencil contains no nodal members but, as appears from the mapping of W_p on W_p^*, it includes instead a certain number of members which are each endowed with d ordinary double varieties which are trajectories of W_p. By the lemma, these make no contribution to the value of $[X_0(W_p)]$; and since both S and S^2 are pseudo-Abelian varieties (of type q), it follows from the inductive hypothesis and equation (2) that $[X_0(W_p)] = 0$.

The result for $k > 0$ now follows from the inductive hypothesis and equation (4), again supposing S to belong to $\{V_q\}$. Evidently, for $k \geqq q$, the system $\{X_k(W_p)\}$ is either generated by trajectories or else has order zero, while for $k < q$, it has order zero.

An immediate corollary of this theorem is that *the canonical invariants of any pseudo-Abelian variety are all zero.* For the intersection number of any set of varieties of appropriate dimensions which belong to a congruence, or have order zero, must be zero.

We consider next the arithmetic genus P_a of W_p, defined by means of the postulation formula (I, 7); from LEFSCHETZ's result for the arithmetic genus of a PICARD variety, and GAETA's formula (I, 11) for the arithmetic genus of a product variety, it at once follows that, in the case $d = 1$, *the arithmetic genus of* W_p *is equal to* $(-1)^{p-1}$. In the case $d > 1$, we may obtain this result by making the assumption that the arithmetic genus is an *enumerative character* of the variety; for on this hypothesis TODD [1] has proved that $P_a + (-1)^p$ is expressible as a homogeneous linear function of the canonical invariants. As previously remarked, TODD's hypothesis has been justified by HIRZEBRUCH [5].

The analytical representation of the pseudo-Abelian varieties, in the case $q = 1$, is effected exactly as in the theory of the elliptic surfaces, and general results are known for all cases, i. e. whether the trajectories are general, harmonic or equianharmonic. But for $q > 1$, the problems of representation of the PICARD varieties, even those of general moduli, which are an essential preliminary, have not so far been considered. However, in the cyclic case, that is, where the involution i_d is cyclic, the form of the analytical representation, for any values of p and q, is readily found (ROTH [33]).

Among the pseudo-Abelian varieties there is an important subclass which belongs to the Abelian varieties considered in § 3; this consists of the superficially irregular improperly Abelian varieties. The justification of this statement depends on a number of results, the first of which is the following:

Every Abelian variety W_p *of superficial irregularity* q $(0 < q < p)$ *and with some plurigenus greater than zero is pseudo-Abelian of type* q.

This is an immediate consequence of equations (3) of § 3, since the involution I_n is clearly invariant under an ∞^q continuous group of transformations of the first kind. The restrictions on the value of q are due to the fact that, from the mapping of W_p on V_p, we deduce that $q \leq p$; and that, if $q = p$, W_p is necessarily a PICARD variety. We next remark that:

With the same hypotheses as before, W_p *contains, in addition to the congruence* $\{V_q\}$ *of trajectories, a congruence* $\{V_{p-q}\}$ *of birationally equivalent varieties* V_{p-q} *which is free from base points and singular points.*

For the congruence $\{V_q\}$ on W_p, which certainly exists, can arise only from a congruence of varieties $\overline{V}_q$ on V_p; hence V_p is special of type q, and therefore contains a complementary congruence of PICARD varieties $\overline{V}_{p-q}$ which in its turn must give rise to the congruence $\{V_{p-q}\}$ on W_p.

It will be observed that this result does not depend on the hypothesis, made in the previous work, that V_q is general. When the latter condition is not satisfied, the congruence $\{V_{p-q}\}$ need not be of PICARD type, though of course it is always Abelian since it is birationally equivalent to an involution on a PICARD variety.

Consider now the mapping of W_p on the d-ple variety W_p^*; since the corresponding varieties V_q^* and V_{p-q}^* are Abelian of genera q and $p - q$ respectively, it follows that the coordinates of the general point P^*

of W_p^* are expressible as rational functions of the coordinates of two points, lying on Abelian varieties of genera q and $p - q$ respectively. Hence the coordinates of the general point P of W_p are expressible as algebraic functions of the coordinates of P^*. Thus (ROTH [34]):

Every Abelian variety W_p of superficial irregularity q $(0 < q < p)$ and with some plurigenus greater than zero is representable parametrically by means of algebraic functions of Abelian functions of genus q and other Abelian functions of genus $p - q$.

In many cases one or both of the congruences $\{V_q\}$ and $\{V_{p-q}\}$ are themselves improperly Abelian, in which case the genera of the Abelian functions required for the parametric representation of W_p can be lowered further.

We should add that the group-theoretic method, on which the above results are based, cannot deal exhaustively with the class of superficially irregular Abelian varieties all of whose plurigenera are zero; such varieties certainly exist – thus, in the case $p = 2$, we have the elliptic scrollar surfaces.

We conclude this section with a brief account of the *para-Abelian varieties*, which bear the same relation to the pseudo-Abelian varieties as the paraelliptic surfaces, introduced by ENRIQUES (a), do to the elliptic surfaces. In order to pass from a pseudo-Abelian variety to a para-Abelian (more precisely, para-pseudo-Abelian) variety W_p of type q, we first replace the congruence of trajectories by a congruence $\{V_q\}$ of PICARD varieties, the irreducible members of which are, as before, non-singular and birationally equivalent; however, this congruence may now contain, in addition to the multiple PICARD varieties of the previous case, a certain aggregate of reducible members of other kinds. We can then prove that W_p possesses a second congruence $\{V_{p-q}\}$ of varieties V_{p-q}, d-secant to $\{V_q\}$, which will likewise contain reducible members. We call the variety W_p para-Abelian if it can be mapped on the d-ple product W_p^* constructed as above, in such a way that the coincidence locus of the representation consists entirely of varieties generated by reducible members of $\{V_q\}$ and $\{V_{p-q}\}$, there being no other exceptional features in the correspondence between W_p and W_p^*.

From the correspondence between W_p and W_p^* we deduce that W_p is in general superficially irregular, and we obtain, as before, a lower limit to the superficial irregularity. We may also prove (ROTH [35]) the following extension of the properties of the canonical systems of a pseudo-Abelian variety: *the canonical varieties $X_k(W_p)$ $(k = p - q,$ $p - q + 1, \ldots)$ either belong to the congruence $\{V_q\}$ or else have order zero.* It is known that, in the case $p = 2$, $q = 1$, this result is best possible; for while the canonical system of a paraelliptic surface is compounded of a pencil of elliptic curves, its SEVERI series has positive order. However, for $p > 2$, the result may in certain circumstances extend to smaller values of k; thus we may show that, *if* $\{V_q\}$ *contains only* ∞^i *reducible members* ($i = 0, 1, \ldots, p - q - 2$), *the above property holds for all* $k \geqq i + 1$.

5. Elliptic and hyperelliptic threefolds. We now give a brief account of the results obtained for the case $p = 3$, in which we are particularly interested.

There are two kinds of pseudo-Abelian threefold, the *elliptic* and the *hyperelliptic*, corresponding respectively to the values $q = 1$, $q = 2$; on the elliptic threefold, the trajectories form a congruence of elliptic curves while, on the hyperelliptic threefold, they form a pencil, rational or irrational, of PICARD surfaces. In the work (ROTH [29, 30]) on these threefolds it is not assumed that the trajectories are general; in all cases, then, we may establish the existence, on the elliptic threefold, of an elliptic pencil of birationally equivalent surfaces and, on the hyperelliptic threefold, of a congruence of birationally equivalent curves; when the trajectories are general, this congruence must be of PICARD type but, when they are special (of type 1) the congruence may be improperly hyperelliptic.

Suppose, first, that W_3 is elliptic; then the congruence Γ of trajectories $\mathfrak{C}$ may contain an infinity of s-fold elliptic curves $\mathfrak{C}_s$ where the number s takes various values, which generate a certain number of irreducible surfaces B_s; or Γ may contain only a finite number of such curves; or again, there may be both surfaces B_s and isolated curves $\mathfrak{C}_s$.

It is clear that any surface belonging to Γ is elliptic (or, in particular, of special PICARD type); applying this consideration to the virtual pure canonical system of W_3, we find that the invariants have the respective values $\Omega_0 = 0$, $\Omega_1 = 1$, $\Omega_2 = -1$, $P_a = 1$. Also from § 4 we see that the superficial irregularity q_2 of W_3 satisfies the equation $q_2 = q + 1$, where q is the irregularity of Γ. Again, denoting by $\{C\}$ the elliptic pencil of surfaces on W_3, we observe that, by virtue of the theory of transformations of elliptic curves, each surface C is invariant under a group G_d of automorphisms which is either cyclic or Abelian of base 2.

In ROTH [29] the case where there are no isolated curves $\mathfrak{C}_s$ is studied in detail; we then have a number of irreducible surfaces B_s each generated by curves $\mathfrak{C}_s$, and it may be shown that each such surface B_s is elliptic of determinant d/s, and such that the pencil of trajectories on it has no multiple members. The curves of Γ cut on C an involution i_d having for $(s-1)$-fold united curves the intersections of the surfaces B_s with C, but with no isolated united points; we may therefore write down the relations between the invariants of Γ and C which follow from the $(1, d)$ correspondence between Γ and C.

Next, suppose that W_3 is hyperelliptic, with a pencil $\{C\}$ of trajectories of genus $\varrho \geqq 0$, and a complementary congruence Γ of curves $\mathfrak{C}$ of genus π. The theory of such a threefold is similar to that of the elliptic surface, or indeed any pseudo-Abelian variety of type 1: thus we have the same relation between the characters ϱ, π and s, the same equivalence*) for the virtual pure canonical system, with the consequence that W_3 has geometric genus ϱ. The numerical invariants are the same as for the elliptic threefold; and the superficial irregularity satisfies the equation $q_2 = \varrho + 2$, or $q_2 = \varrho + 1$, according as Γ is properly or improperly hyperelliptic.

From the formulae for the plurigenera we find, as in the theory of surfaces, that a necessary and sufficient condition that $\pi = 0$ is that $P_{12} = 0$. We also find (ROTH [30]) that when $\pi = 0$, the congruence Γ always possesses unisecant surfaces, so that W_3 is birationally equivalent to a point-cone projecting a surface image of Γ.

The improperly Abelian threefolds have been studied in ROTH [31], where the principal types have been determined by geometrical considerations based on the known classification of the hyperelliptic surfaces; in order to determine the orders of the corresponding involutions on the PICARD threefolds, one must have recourse to the group-theoretic method mentioned in § 4, but it has been shown by constructing examples that all the types which are *a priori* possible actually exist. All these threefolds are generated by congruences of elliptic curves which are trajectories in the elliptic case and complementary congruences in the hyperelliptic case.

In the same work the para-Abelian threefolds of type 1 are also considered; such threefolds contain congruences of elliptic curves whose irreducible members are all birationally equivalent: hence, in particular, the virtual pure canonical system is in each case para-elliptic, and the numerical invariants are given by $\Omega_0 = 0$, $\Omega_1 = 1$, though in general $\Omega_2 \neq -1$. It is also found that the superficial irregularity of such a threefold is not less than the irregularity of the associated congruence of elliptic curves.

6. Threefolds which possess finite continuous groups of automorphisms. In previous sections we have encountered various classes of threefold endowed with finite continuous groups of automorphisms; the complete discussion of the problem is due to HALL [2], whose work forms the basis for §§ 6, 8.

We first remark that, if an algebraic variety is invariant under such a group, it must be invariant under an *algebraic* group of the same dimension; and, further, that it is always possible to find an algebraic subgroup of this group which is *permutable*. These results are given by PAINLEVÉ (a) for the special case of surfaces, but they apply to any variety whatever, and will be assumed in what follows.

*) Strictly speaking, when Γ is improperly hyperelliptic the equivalence formula contains an additional term representing a virtual surface of order zero.

I. *One-parameter groups.* If a threefold V is invariant under a one-parameter group, the trajectories of the group form a congruence of rational or elliptic curves. In the latter case V is elliptic (§ 5); in the former, as we shall show, V is *scrollar*, i. e. birationally equivalent to a line-ruled threefold. To prove this result we require to determine a unisecant surface to the congruence, Γ, say (IV, 4).

Now the ∞^1 group of transformations set up on the generic curve of the congruence either consists of all the collineations with two fixed united points or else all the elations with one fixed united point. In the second case the locus of this united point gives the required unisecant; in the first case the locus of united points is a bisecant surface. We shall show that it must consist of two unisecant surfaces; this follows from a result of PAINLEVÉ (a) that, if a surface is invariant under a rational ∞^1 group of automorphisms, the locus of united points (assumed distinct) consists of two unisecants to the trajectories*).

Let Γ be mapped by the points of a surface F; to a curve on F corresponds a surface of V belonging to Γ; by PAINLEVÉ's result, the locus of united points on this surface consists of two curves, which lie on the surface U, say, locus of united points on V. Now let $|\mathfrak{C}|$ be a linear system, of freedom $r > 1$, of irreducible curves of positive grade on F; to this corresponds a rational system of surfaces on V. Since r general points of V determine r curves of Γ and so r points of F; and since these points determine a unique curve of $|\mathfrak{C}|$, to which corresponds a unique surface of the rational system passing through the r chosen points of V, it follows that this system has index 1 and is therefore linear. The system meets U in a linear system of curves, each member of which breaks up into two variable components. Now since the system has positive grade, it cannot be compounded of a pencil; and since, by BERTINI's theorem, such a situation is impossible on an irreducible surface, we conclude that U consists of two unisecant surfaces, and hence that V is scrollar.

II. *Two-parameter groups.* In this case the trajectories on V may form a congruence of rational curves or a pencil of surfaces which may evidently be rational, hyperelliptic (of rank 1) or elliptic scrollar; for if the surfaces were scrollar of higher genus the trajectories would consist of the aggregate of rational curves contained in them, so that we should again have a congruence of rational curves.

If the trajectories are hyperelliptic surfaces, V is hyperelliptic; if they are elliptic scrollar, the transformations which leave their generators invariant form an ∞^1 subgroup, so that in this case V is a particular type of scrollar threefold. *If the trajectories are rational surfaces, then V is planar.*

*) It may be noted that this result (obtained by transcendental methods) was established before ENRIQUES's extension of NOETHER's theorem (IV, 7).

In order to prove this result we first determine an ∞^1 rational subgroup of the given group. Consider the given group as operating on any particular trajectory A; since A is regular, it follows from § 2 that we can find a birational transform A' of A such that the corresponding transformations of A' are projective. Now, given any ∞^2 group, we can determine from it many ∞^1 subgroups; further, if the group is both algebraic and projective, a subgroup may be found which is also algebraic. This follows from the fact that any ∞^2 group is integrable (for a non-integrable group must contain an ∞^3 simple subgroup) and that the integrable projective groups have all been classified (§ 1).

When we transfer back to A we now have an ∞^1 rational group operating on A. If each transformation of A arises from one transformation of the original group for the entire threefold V*), we at once obtain the desired rational subgroup. However, it may happen that a given transformation of A arises from a finite number $n > 1$ of transformations of V; this will occur if there is some transformation, other than identity, leaving A totally invariant. In this case the transformations of the original group which give rise to the ∞^1 group on A form an algebraic group which may be reducible. In any case this group consists of m $(1 \leqq m \leqq n)$ components; each component has dimension 1, and one (at least) must contain the identity and will be a continuous algebraic group (see ENRIQUES-CHISINI, a: the transformations there considered are collineations, but this fact does not enter into the proof). Since the trajectories on A of this continuous group are algebraic curves, it follows that the group must itself be rational (CASTELNUOVO-ENRIQUES [3]). Thus in every case we have the required subgroup.

On each surface of the pencil $|A|$ of trajectories we thus have a pencil of rational curves, namely the trajectories of this subgroup; we may use these to transform V birationally to a cone V' on which $|A|$ is mapped by a pencil of cones. The generic surface section of V' thus contains a pencil of rational curves, and so is scrollar; and we can now transform V' birationally to a cone V'' generated by ∞^1 planes.

III. *Three-parameter groups.* In this case, if the group is intransitive, the trajectories on V may be rational curves or rational or elliptic scrollar surfaces; in all three cases previous methods show that V is then either scrollar or planar. Next, supposing that the group is transitive but not permutable, we can always obtain an ∞^1 or ∞^2 permutable subgroup to which the former considerations will apply. Again, if the group is permutable and completely transitive, V must be a PICARD

*) This will always be the case when the trajectories are elliptic curves or PICARD surfaces; any other situation can arise only when there are rational trajectories.

threefold; while if it is only generally transitive, V is quasi-Abelian (§ 3).

In conclusion, then, we have the following results:

A threefold which is invariant under a continuous algebraic ∞^1 *group of automorphisms is scrollar or elliptic. A threefold which is invariant under a continuous* ∞^2 *group of automorphisms is scrollar, planar or hyperelliptic. A threefold which is invariant under a continuous* ∞^3 *group of automorphisms is Abelian (of rank* 1*) or quasi-Abelian provided that the group is permutable and transitive: otherwise it belongs to one or other of the previous classes.*

It is of course to be understood that the scrollar or planar threefolds mentioned above may in particular be birational; that they cannot be unirational is shown by the following theorem (FANO [10]):

A completely regular threefold which admits a finite continuous group of automorphisms is either birational or is not even unirational.

For since the threefold V is completely regular, we can, by § 2, determine on it an algebraic congruence of rational curves $\mathfrak{C}$ with unisecant surface C, which means that V is expressible as a product $V = \mathfrak{C} \times C$. It then follows from the formulae of I, 11 that V is completely regular if, and only if, C is regular of genus zero. If C is rational, V is obviously birational; but if C is irrational, FANO's theorem of IV, 4 shows that V cannot be unirational.

7. Extensions to varieties of higher dimension. As soon as we seek to extend the previous investigation to the study of the varieties V_n $(n > 3)$ which admit continuous groups of automorphisms, we are faced with the difficulty that the groups in question may be non-integrable: in that case it would seem that our methods do not lead to precise results. However, it is known that every *permutable* group is integrable; we shall therefore limit the discussion to groups of this kind.

Consider, in the first place, a variety V_n $(n > 3)$ endowed with a permutable (generally) transitive continuous group of ∞^n automorphisms; such a variety is called quasi-Abelian (§ 3), and for it SEVERI (b) has established the following results:

Every superficially regular quasi-Abelian variety V_n *is birational. In all cases the superficial irregularity* q *of* V_n *satisfies the inequality* $q \leqq n$; *and* V_n *is birationally equivalent to the product of a* PICARD *variety* V_q *and a linear space* S_{n-q} $(0 \leqq q \leqq n)$. *Conversely, any such product is a quasi-Abelian variety of superficial irregularity* q.

The demonstrations given by SEVERI are transcendental and somewhat lengthy; in the original version (b) they depend on an unproved hypothesis concerning the possibility of eliminating certain exceptional varieties from V_n, though in SEVERI [12] this hypothesis has been dispensed with. It would, however, be very desirable to have a geometrical proof of these theorems; such a proof could presumably be based on the known properties of PICARD varieties. In any case the results show that *every quasi-Abelian variety of superficial irregularity* q (> 0) *is merely a special case of a pseudo-Abelian variety of type* q *and determinant unity*. It follows that the two possible ways, suggested in § 3, of extending the notion of PICARD variety in effect reduce to one, leading to the pseudo-Abelian variety in every case.

Having established the above properties of quasi-Abelian varieties, we are in a position to study the varieties V_n which admit permutable continuous groups of ∞^r automorphisms, where $r < n$, or $r > n$. To begin with, we observe that SEVERI's theorem holds equally in the case where a variety V_n admits a transitive group of dimension $n' > n$; for his proof is based on the consideration of n linearly independent simple integrals of the third kind on V_n which exist by virtue of the group property, and these are obtained *a fortiori* when $n' > n$. It is then possible to choose n of these integrals, after which the proof proceeds as before.

As already remarked, there is no loss of generality in assuming that the ∞^r group ($r < n$) under which V_n is invariant is algebraic; we then know that the trajectories of the group must be either PICARD varieties or quasi-Abelian (in particular, birational) varieties of some dimension $r' \leqq r$. When they are PICARD varieties, $r' = r$; the group then operates in a simply transitive manner on the general trajectory, amd V_n is a pseudo-Abelian variety of type r. We now prove (HALL [3]) that, *in every other case*, V_n *is birationally equivalent to the product of a linear space* S_d *and a variety* V_{n-d} *which, for* $r' > d$, *is pseudo-Abelian of type* $r' - d$.

To this end we fix our attention on any simple irreducible trajectory A. The group determined on A has dimension r'' ($r' \leqq r'' \leqq r$); consider first the case where $r'' = r$, so that each transformation of A arises from a finite number of transformations of V_n. By SEVERI's theorem, A is birationally equivalent to the product A', say, of a PICARD variety V_p and a space $S_{r'-p}$. Regarding the group as operating on A', we see that it contains a subgroup corresponding to those transformations which determine the identical transformation on V_p. This subgroup acts on $S_{r'-p}$, and is thus equivalent to a projective group whence, as we know, it contains a rational ∞^1 subgroup.

Now the corresponding transformations of A arise from an ∞^1 algebraic group for V_n. If this is reducible, it consists of a finite number of components, each of dimension unity and of these, one at least is a finite continuous group, determining a congruence of rational trajectories on V_n. In order to transform V_n into a cone whose generators correspond to these trajectories, we require to construct a variety unisecant to these curves. The existence of such a variety is established as in § 6, reasoning inductively from $n - 1$ to n.

Since, by hypothesis, the given group for V_n is permutable, every subgroup, in particular the above ∞^1 subgroup, is invariant. This implies that the congruence of generators of the cone is invariant under the entire group, any transformation of which either leaves every curve invariant or else determines a birational transformation acting on the curves as elements. Hence the general prime section of the cone is invariant under a group induced by that group which acts on the generators, and the trajectories of the former will have dimension $r' - 1$.

Evidently the previous considerations may now be applied in turn to this prime section. We continue to reason thus for r' stages in succession, unless, at some intermediate stage, on taking a prime section we obtain a pseudo-Abelian variety. Our result is thus established in the case $r'' = r$.

In the case $r'' < r$, an ∞^1 group on a trajectory no longer arises from an ∞^1 group for V. In this case the identical transformation on A arises from $\infty^{r-r''}$ transformations of V_n; these form an algebraic group which either is or contains a continuous algebraic group (of dimension $r - r''$). The trajectories of this group constitute a congruence of which the original trajectories are compounded; in particular, the two sets of trajectories may be the same.

These new trajectories are quasi-Abelian, not PICARD, varieties, since those which lie on A are totally invariant for the group. If the group determined on one of the trajectories has dimension $r - r''$, we can, as before, determine an ∞^1 subgroup for V_n, and this will be an invariant subgroup for the entire group of dimension r.

On the other hand, if the group on a trajectory has dimension less than $r - r''$, we can proceed to obtain a group of lower dimension. If this occurs at each stage of the process of reduction, we shall eventually obtain a congruence of curves invariant under a group of dimension greater than unity. Since these curves must be rational, it follows that, by imposing a suitable number of invariant points on V_n, we obtain, as before, a rational ∞^1 group. Thus the result is established in all cases.

8. Other types of automorphism. We conclude this chapter with brief accounts of threefolds which admit other species of automorphism than continuous groups. We consider first the problem of constructing threefolds which are invariant under a *continuous series* of automorphisms but not under any finite continuous group: while there are no algebraic curves or surfaces with this property, FANO's theorem (§ 6) shows that, for threefolds, the question is significant.

Let V be a threefold endowed with such a series of automorphisms; then the transforms of the general point of V may lie on an algebraic curve or an algebraic surface or may not be restricted to any subvariety. In the first case the curve in question may *a priori* be rational or elliptic: but that it cannot be elliptic is seen as follows (HALL [2]). Supposing the curve to be elliptic, we have on V a congruence of elliptic curves, and any surface belonging to the congruence must be elliptic. On this surface, a transformation of the first kind, given on one of the curves, determines a unique transformation, and hence a unique transformation on each curve of the pencil on this surface. Now consider the generic curve $\mathfrak{C}$ of the congruence and also a pencil of (elliptic) surfaces belonging to the congruence and each containing $\mathfrak{C}$. Every transformation of the elliptic group on $\mathfrak{C}$ determines a unique transformation of the other curves of the congruence, so that V is invariant under a continuous ∞^1 group, contrary to hypothesis.

It follows, then, that the congruence on V consists of *rational* curves; we thus require that these curves should not be transformable to lines – provided always that a finite continuous group of automorphisms of V does not arise in some way unconnected with the congruence. For the curves may always be transformed to conics (IV, 4) which generate a birational transform V' of V; and a prime of the ambient space of V' determines a pair of points on the generic conic of the congruence, and so an involutory transformation having these for double points; such involutions and their products give the required continuous series of automorphisms. An illustration is furnished by the ENRIQUES primal (V,9).

If instead the transforms of a point are restricted to a surface, the latter must *a priori* be rational, scrollar, elliptic or hyperelliptic (of rank 1). Now the scrollar and elliptic surfaces give congruences of invariant curves, as in the case already considered; so do the elliptic scrollar surfaces, since those transformations which leave their generators invariant must form a continuous series. There remain the cases of the rational and hyperelliptic surfaces; the existence of threefolds of the required type containing pencils of such surfaces is suggested, but not effectively established, by the following considerations (HALL [2]).

In an attempt to construct a threefold on which the transforms of a point are necessarily restricted to a rational surface, we appeal to BALDASSARRI's result (BALDASSARRI [3]) that, on a primal V_3^m of S_4 with an $(m-3)$-ple plane, there exists a system of curves unisecant to the cubic surfaces of the pencil residual to the multiple plane. Each such curve determines an involutory transformation by projection from the points where it meets each cubic surface, so that we obtain a continuous series of such transformations. Unless the threefold in question is planar, and so birational, we thus construct a model which is not invariant under a finite continuous group of automorphisms.

In the second case the obvious type of threefold is one for which the surfaces of the pencil are of varying moduli and possess a continuous system of unisecant curves; the two points determined on a surface by any two unisecants taken in one order or the other would give a unique transformation of the first kind. More generally, the curves could be plurisecant, provided the parameter-sums for the two sets were not equivalent (cf. ENRIQUES [9]).

Turning to types of threefold for which the transforms of a point invade the variety, we remark that the non-singular cubic primal of S_4 certainly admits continuous series of transformations, e. g. those obtained as products of projections from the points of the primal; to show that there exist no finite continuous groups of automorphisms would be equivalent to establishing the irrationality of the primal – as has been pointed out by SNYDER [1], who has constructed various series of transformations.

We now pass to threefolds which admit *infinite discontinuous groups* of automorphisms but not, in general, continuous series of transformations. In the analogous theory of surfaces we have the result, due to ENRIQUES [9], that any surface which is endowed with such a group must either contain a pencil of elliptic curves or else be regular with genera and plurigenera unity. For the surfaces of the first type, the construction of the group has been given by ENRIQUES; further contributions to the theory will be found in GAETA [1]. For surfaces of the second type, however, no general rule of construction can be given, and each case must be considered on its own merits.

There is as yet no comparable theory for threefolds; in fact, the proof of ENRIQUES's theorem is based on the remark that, for all surfaces having the stated property which are not of the second type, it is possible to construct a simple canonical or pluricanonical model. Since the analogous result does not hold for threefolds, we can hardly expect to obtain such precise information concerning them. Nevertheless, to begin with, a few examples readily suggest themselves: thus, among threefolds which contain congruences of elliptic curves and which admit such groups of automorphisms we have the quartic primal of S_4 with two nodes, and various other threefolds discussed in ROTH [15]. A different type is exemplified by the product of a curve of genus greater than unity and any surface which contains a pencil of elliptic curves and which is invariant under a discontinuous group.

When the threefold in question contains a pencil of invariant surfaces which do not belong to a congruence of elliptic curves, these surfaces may be rational or hyperelliptic or surfaces, with genera and plurigenera unity, which are invariant under a discontinuous group. An example of the last kind is furnished by the product of such a surface and a curve of genus greater than unity.

The case where the transforms of a given point are unrestricted is illustrated by the quartic primal of S_4 which has a finite number (at least four) of nodes not lying in the same plane; other examples, depending on methods suggested by the theory of elliptic functions, may be constructed by analogy with various types of surface given by PAINLEVÉ (HALL [2]), and more interesting illustrations may be found by appealing to the theory of hyperelliptic functions.

Appendix.

The following Appendix is intended to serve as a brief guide to the principal concepts, notations and results which have been employed in this book. No detailed proofs and no references are given; but so far as the invariantive geometry of curves and surfaces is concerned, we refer the reader to the appropriate works listed in the Bibliography, while for the projective geometry any standard text may be consulted.

1. Complex projective geometry. The *homogeneous coordinates* of a point in complex projective space S_r, or $[r]$ of r dimensions are an ordered set (or coordinate vector) $(x_0, x_1, \ldots, x_r)$ of $r+1$ complex numbers, not all zero; and any two sets whose corresponding elements are proportional determine the same point. Projective geometry deals with those properties of geometrical configurations which are invariant under the group of *collineations* or *projective transformations*; these are of the form

$$\varrho x_i' = a_{i0} x_0 + a_{i1} x_1 + \cdots + a_{ir} x_r \quad (i = 0, 1, \ldots, r), \tag{1}$$

where ϱ is a factor of proportionality and the determinant $|a_{ij}| \neq 0$.

In this geometry we are concerned only with systems of homogeneous polynomial equations; but it is sometimes convenient to take instead the ratios x_i/x_0 $(i = 1, 2, \ldots, r)$ for *non-homogeneous coordinates*. In that case, so as to avoid exceptional elements, a class of fictitious or *improper* points is then introduced, namely those points for which $x_0 = 0$.

2. Linear spaces. Projection. The totality of points whose coordinates satisfy the equation

$$a_0 x_0 + a_1 x_1 + \cdots + a_r x_r = 0 \tag{2}$$

is called a *prime*, and denoted by S_{r-1} or $[r-1]$. And generally, the *intersection* of, or totality of ∞^k points common to, $r-k$ linearly independent primes $(0 \leq k \leq r-1)$, is called a *k-dimensional linear space*, and denoted by S_k or $[k]$. In particular, for $k = 0, 1, 2$, we have a point, line and plane respectively. It follows from this definition that a space S_k is uniquely determined by any $k+1$ given points whose coordinate vectors are linearly independent. The parametric equations of S_k may be written in the form

$$\varrho x_i = a_{i0} \lambda_0 + a_{i1} \lambda_1 + \cdots + a_{ir} \lambda_r \quad (i = 0, 1, \ldots, r), \tag{3}$$

where $\lambda_0, \lambda_1, \ldots, \lambda_r$ are homogeneous parameters.

An important role in this geometry is played by the process known as *projection*. We define the projection of a point P of S_r from a *vertex* S_h $(0 \leq h \leq r-2)$ on to a space S_{r-h-1} as the intersection of that space with the S_{h+1} which passes through P and the vertex. This notion is further developed below.

3. Primal. A *primal* is the locus of points whose coordinates satisfy an equation of the form

$$f(x_0, x_1, \ldots, x_r) = 0, \tag{4}$$

where f is a homogeneous polynomial of some order n. The primal is said to be of *order* n, and is denoted by V^n_{r-1} or W^n_{r-1}. In the case $n=2$, it is called a *quadric*. V^n_{r-1} is said to be *irreducible* or *reducible* according as f is irreducible or reducible.

4. Generic point. In algebraic geometry the concepts *generic* and *general* frequently occur, and are of great importance. As we shall now exemplify, they may be interpreted in several ways, according to the context in which they arise.

I. Consider a given (algebraic) condition c, imposed on the points of V^n_{r-1} and not satisfied by all the points of the primal. It may happen that, in a certain problem, we wish to exclude those points of V^n_{r-1} which satisfy c: in that case all the remaining points of V are called *generic* or *general*.

II. While this concept is relative (to the problem in question) in the case where V^n_{r-1} is *irreducible* it may be rendered absolute. Thus we envisage a point whose coordinates are indeterminates, and which cannot therefore satisfy any condition not satisfied by all points of V^n_{r-1}. For this to be possible it is necessary that the polynomial f in (4) should not break up into factors, each of which vanishes on part but not the whole of V^n_{r-1}; that is, V^n_{r-1} must be irreducible. The condition of irreducibility is also sufficient to secure the existence of the generic point.

III. The above notion extends in an obvious way, as the following examples, which are self-explanatory, suggest.

a) A generic line meets V^n_{r-1} in n distinct points. [This follows from (3) and (4).]

b) In $[r]$, a $[k]$ and a $[r-k]$ in general meet in a single point (§ 2).

c) The projection of a point from a given vertex is in general a point.

d) The *tangent* lines at the generic point P of V^n_{r-1} (i. e. lines having at least two coincident intersections there with V^n_{r-1}) lie in a prime, called the *tangent prime* at P. A point of V^n_{r-1} is called *singular* if, and only if, the tangent prime there is indeterminate, i. e. if, and only if, all the derivatives $\partial f/\partial x_i$ vanish at the point in question.

5. Algebraic variety. The aggregate of points common to a finite number of given primals of S_r is called an *algebraic variety*, or *manifold* V. If, for every i ($i = 1, 2, \ldots, r-k-1$) there exists some S_i not meeting V while every S_{r-k} meets V (i. e. has points in common with V) we say that V has *dimension* k.

Let S_{r-k-2} and S_k be two spaces in general position with respect to V and each other; and let W be the *projection* of V from S_{r-k-2} on S_{k+1} (i. e. the locus of the projection P' of a point P of V). If W is a primal (possibly reducible) we say that V is *pure*. If, further, W is irreducible, then V is likewise irreducible. Also the order n of W, which is equal to the number of points in which V meets a generic S_{r-k}, is called the order of V; and we denote V by the symbol V^n_k. For $k = 1, 2, 3$, we have a *curve*, *surface* and *threefold* respectively.

Unless the contrary is stated we shall assume that V is irreducible; to such a variety we may apply the concept of *generic point* (§ 4). Then, at the generic point P of V there exists a unique space S_k, corresponding to the tangent prime to W at P'; this is called the tangent space S_k to V at P. A point of V is called *singular* (or multiple) if, and only if, the tangent space there is indeterminate. A variety which is free from singular points is called *non-singular*.

6. Projective characters. The order n of a variety is the simplest example of a *projective character*, i. e. a character which is invariant under a general projection. For a curve in S_r $(r \geqq 2)$ we have also to consider the *rank*, i. e. the number of its tangents which meet a given S_{r-2} in general position. For a surface in S_r $(r \geqq 4)$ there are four important projective characters, namely the order and rank of a generic *curve section* (section by a prime); the *class*, or number of tangent planes which meet a generic S_{r-2} in a line; and the *ceto*, or number of tangent planes which meet a generic S_{r-4} (in a point).

7. Some particular classes of variety. The following types of variety are specially important in our work.

I. SEGRE *varieties*. Let S, S' be any two given spaces of arbitrary dimensions: then the aggregate of unordered pairs (P, P'), where P, P' describe S, S' respectively is called a SEGRE variety, or the *product* of S, S', and is written as $S \times S'$. This concept at once extends to the product of any number of given spaces.

II. *Product varieties*. If V, V' are any two irreducible varieties, the product $V \times V'$ may be defined as above; and the definition extends to the product of any number of given varieties of assigned dimensions.

III. *Grassmannians*. In $[r]$ a space $[k]$ is uniquely determined by $k+1$ generic points lying in it. From the matrix formed by the coordinates of these points we may extract $\binom{r+1}{k+1}$ determinants of order $k+1$, which we call the GRASSMANN coordinates of $[k]$. Taking these to be homogeneous coordinates of a point of $[R]$, where $R = \binom{r+1}{k+1} - 1$, we obtain a representation of the spaces $[k]$ on the points of a variety $G(k, r)$, called the Grassmannian of the spaces in question. The simplest significant example, furnished by the case $k=1$, $r=3$, is a quadric of $[5]$.

8. Birational geometry. Let P, P' be generic points of two spaces S_r, S'_r whose coordinates $(x_0, x_1, \ldots, x_r)$, $(x'_0, x'_1, \ldots, x'_r)$ are connected by the relations

$$\varrho x'_i = f_i(x_0, x_1, \ldots, x_r) \quad (i = 0, 1, \ldots, r), \tag{5}$$

where the functions f_i are polynomials of the same order. The equations (5) define a *unirational* representation of S'_r upon S_r, or transformation of S'_r into S_r; to a point P there corresponds a unique point P', but to a point P' there corresponds a set of n distinct points P, where in general $n > 1$. If, however, $n = 1$, the equations are rationally invertible: in that case we say that they represent a *birational* (CREMONA) transformation of S'_r into S_r (or S_r into S'_r).

It may happen that, while (5) are not in general rationally invertible, they are so invertible provided that P describes a certain variety V_k; in that case P' will describe a second variety V'_k : V_k, V'_k are then said to be *birational transforms* of each other, and are *birationally equivalent*. In particular, V'_k may coincide with V_k, in which case we have a birational transformation (or *automorphism*) of V_k into itself.

Birational geometry studies those properties of algebraic varieties which are invariant under birational transformation; for the purpose of this study any particular member of a class of birationally equivalent varieties may be selected, for example, some projective model. It is for this reason that projective geometry plays an important part in the development of birational geometry.

An interesting class of manifold consists of those varieties V_k which are birationally equivalent to a space S_k; such varieties are called *birational*; evidently they admit a parametric representation of the form

$$\varrho x_i = f_i(\lambda_1, \lambda_2, \ldots, \lambda_k) \quad (i = 0, 1, \ldots, r), \tag{6}$$

where $\lambda_1, \lambda_2, \ldots, \lambda_k$ are *essential* (non-homogeneous) parameters, and the f_i are rational functions such that the equations (6) are rationally invertible. More generally, we may envisage a variety V_k which admits a rational parametric representation by means of equations which are no longer rationally invertible; such a variety is called *unirational*. This concept is significant only for $k \geqq 3$, for it is known that *any curve or surface which is unirational is also birational*; it is therefore usual to speak of a *rational* curve or surface without reference to the particular representation in question.

It should be noted that the correspondence between V_k and V'_k, even when it is birational, will in general possess exceptional features. Consider, for instance, the process of projection, which of course is a particular kind of birational transformation. When a curve V_1 of S_3 is projected from a point O on to a plane, the projection V'_1 will possess a certain number of double points each of which arises from a pair of points of V_1 which are collinear with O. A different type of exception presents itself when a surface V_2 of S_4 is projected on to S_3 from a simple point O of itself; in that case the point O has no corresponding point on the projection V'_2, but the *neighbourhood* of O, consisting of the aggregate of tangent directions to V_2 at O, corresponds to a line on V'_2.

Ex. All Segre varieties and all Grassmannians are birational; but their representations upon the corresponding linear spaces in general possess exceptional features.

9. Geometry on a curve. In the case $k = 1$, the birational geometry, or *geometry on a curve*, is notably simple and the results accordingly complete. A first simplifying factor in the situation is the result that *any curve can be birationally transformed into a non-singular curve of* S_R ($R \geqq 3$). The projection of this curve on to a plane from a generic vertex will possess a finite number of *nodes*, or double points at each of which there are two distinct branches (§ 8). A plane curve of this kind, not necessarily a projection, is said to possess *ordinary singularities*. In all that follows we assume that the curves under discussion are either non-singular or are plane curves with ordinary singularities.

For plane curves which possess singularities of a higher type, e. g. cusps or triple points with, possibly, coincident branches, there exists a systematic theory by which they may be resolved, by successive birational transformations of the curve, into sets of singular points of lower multiplicities and, ultimately, into nodes. In this theory an important concept is that of *proximate point*, according to which a singular point of complicated type may be conventionally regarded as a succession of singular points of simpler kinds.

The geometry on a curve is based on the notion of *linear series* of sets of points which we now introduce. Let $\varphi_i = 0$ ($i = 0, 1, \ldots, r$) be the equations of $r+1$ linearly independent primals of the same order, situated in S_R ($R \geqq 2$); then the aggregate of primals represented by the equation

$$\lambda_0 \varphi_0 + \lambda_1 \varphi_1 + \cdots + \lambda_r \varphi_r = 0\,, \tag{7}$$

where $\lambda_0, \lambda_1, \ldots, \lambda_r$ are parameters, is called a *linear system* of *dimension* or *freedom* r; in the cases $r = 1, 2, 3$, we have a *linear* (or *rational*) *pencil*, *net* and *web* respectively. This system cuts the curve $\mathfrak{C}$, also assumed to be in S_R, in a corresponding aggregate of sets of points which is termed a *linear series*. The simplest example of such a series is provided by that cut on $\mathfrak{C}$ by the primes of S_R.

The system (7) may have a number of fixed, or *base*, points on $\mathfrak{C}$; it is then usually convenient to consider the sets of the series as consisting only of variable points. In any case, the number of points in the generic set is termed the *order* of the series.

The importance of this concept lies in the fact that, *in any birational transformation of a curve, linear series transform into linear series.* For, if $\mathfrak{C}'$ is the birational transform of $\mathfrak{C}$ by the equations (5), the linear system (7) transforms to another linear system, so that any linear series on $\mathfrak{C}$ transforms to a like series on $\mathfrak{C}'$; and since the equations (5) are rationally invertible on $\mathfrak{C}'$, any linear series on $\mathfrak{C}'$ will also transform to a linear series on $\mathfrak{C}$.

The above result allows us to substitute for $\mathfrak{C}$ any convenient model of a class of birationally equivalent curves; we shall often take this to be a plane curve with ordinary singularities.

The theory of linear series may also be formulated in terms of function theory. Let φ_0, φ_1 be any two polynomials of the same order in the coordinates which do not vanish identically on $\mathfrak{C}$; then the ratio $\varphi_1 : \varphi_0$ assumes a unique finite value at the generic point of $\mathfrak{C}$, and is said to be a *rational function* of a point of $\mathfrak{C}$. The function $\mathfrak{R}$ in question has the same number of zeros as it has poles, both counted with the proper multiplicities. If λ is any given constant, the set of zeros of the function $\mathfrak{R} - \lambda$ is called a *set of constant level* of $\mathfrak{C}$; evidently the set is cut on $\mathfrak{C}$, residually (possibly) to a number of fixed points, by the linear pencil $\varphi_1 - \lambda\varphi_0 = 0$, of primals. And, generally, with an obvious extension of the notation, any linear series can be regarded as the aggregate of all sets of constant level of rational functions of the form

$$\lambda_1 \mathfrak{R}_1 + \lambda_2 \mathfrak{R}_2 + \cdots + \lambda_r \mathfrak{R}_r\,.$$

10. Adjoint curves and complete series. Let $\mathfrak{C}$ be a plane curve with ordinary singularities, say d nodes; then any curve which passes simply through all the nodes is termed *adjoint* to $\mathfrak{C}$. Evidently all adjoint curves of assigned order form a linear system, namely the subsystem of plane curves of that order which are constrained to pass through the d nodes.

The significance of the adjoints resides in the two theorems:

a) Any linear series on $\mathfrak{C}$ may be cut out by adjoint curves of some order residually (possibly) to a set of fixed points on $\mathfrak{C}$.

b) The adjoint curves of any assigned order cut a *complete* series on i. e. one not contained in any more ample linear series of the same order.

The concept of complete linear series has an interesting counterpart in projective geometry. Suppose that a curve in S_R has the property that the primes of S_R cut a complete linear series on it: this means that the curve cannot be obtained as the projection of another curve of the same order lying in some space of higher dimension. Such a curve is said to be *normal* in S_R.

It follows from b) that every linear series containing a given set of points of $\mathfrak{C}$ belongs to a *unique complete* linear series containing that set. Such a set may have freedom zero, i. e. the set may be *isolated* on $\mathfrak{C}$.

A linear series of order n and *freedom* (or dimension) r is denoted by the symbol g_n^r. Evidently just one set of the series can be made to contain r generic points of $\mathfrak{C}$.

Ex. If $\mathfrak{C}$ is a non-singular plane cubic, every point of $\mathfrak{C}$ is isolated, for otherwise $\mathfrak{C}$ would contain a g_1^1 and so be rational (cf. § 14).

11. Linear equivalence. The notion which we now introduce is fundamental in algebraic geometry. We say that any two sets α, β of points on $\mathfrak{C}$ are *linearly equivalent* if they are both contained in a linear series on $\mathfrak{C}$: in that case we write $\alpha \equiv \beta$. In other words, $\alpha \equiv \beta$ if, and only if, after subtraction of any common points, α and β are the respective sets of zeros and poles of a rational function on $\mathfrak{C}$.

The relation of linear equivalence between sets is symmetrical, reflexive, transitive and additive. Of these properties, the first three are obvious, while the last follows from the remark that the product of any two rational functions on $\mathfrak{C}$ is itself a rational function on $\mathfrak{C}$.

A linear series of which α is a typical set is denoted by $|\alpha|$; usually this symbol denotes the (unique) complete series defined by α. Similarly, the symbol $|\alpha + \beta|$ usually denotes the complete linear series defined by the set $\alpha + \beta$, which is the same as that obtained by adding a set of $|\alpha|$ to a set of $|\beta|$; in particular, we may have $\beta = \alpha$, in which case the series is denoted by $|2\alpha|$; and similarly for any multiple of $|\alpha|$.

12. Series of equivalence. As regards the subtraction of sets, it is clear that the set $\alpha - \beta$ effectively exists only if β is contained in α; thus in particular the set $\alpha - \alpha$, which we call the *null set*, is effective. We now define a *virtual set* as any difference $\alpha - \beta$ of two effective sets on $\mathfrak{C}$, with the convention that any two differences $\alpha - \beta$ and $\alpha' - \beta'$ define the same virtual set if $\alpha + \beta' = \alpha' + \beta$, i. e. if the sets $\alpha + \beta'$ and $\alpha' + \beta$ are identical.

In the field of virtual sets as so defined, for every effective set α there exists a set $-\alpha$, namely the difference between the (unique) null set and the set α.

Addition and subtraction of virtual sets are defined by obvious rules which we need not state; while a linear equivalence of the form $\alpha - \beta \equiv \gamma - \delta$ holds if, and only if, in the field of effective sets we have $\alpha + \delta \equiv \beta + \gamma$.

The order of any virtual set $\alpha - \beta$ is the difference between the orders of α and β.

The transition from effective to virtual sets, which is analogous to the introduction of the negative integers in arithmetic, leads at once to the concept of a *series of equivalence*, i. e. an aggregate of mutually equivalent virtual sets which may possibly include a totality of effective sets forming a complete linear series. Such a concept is a powerful unifying force in the theory.

13. Jacobian series. The canonical series. Consider, on a plane curve $\mathfrak{C}$, a series g_n^1 of which α is a typical set. A *double point* of g_n^1 is defined as a point of α which counts twice in the set of the series to which it belongs; and the set of all double points of the g_n^1 is termed the *Jacobian set* of g_n^1, and denoted by α_j.

Ex. In the case where g_n^1 is cut on $\mathfrak{C}$ by the pencil of lines through a general point P, a double point is a point of contact of a tangent to from P, and α_j is the set of all such points of contact, lying on the first polar of P.

The Jacobian sets enjoy the following properties:

a) The Jacobian sets of all series g_n^1 contained in a given g_n^r $(r \geqq 2)$ are mutually equivalent.

b) A simple base point of a g_n^1 counts twice as a member of α_j.

By virtue of a) we may envisage the *Jacobian series* $|\alpha_j|$ as the complete linear series of Jacobian sets of series g_n^1 in g_n^r. And it follows from b) that, if $|\alpha|$ and $|\beta|$ are any two linear series of positive freedom, the Jacobian sets of the series $|\alpha + \beta|$ satisfy the equivalences

$$(\alpha + \beta)_j \equiv \alpha_j + 2\,\beta \equiv \beta_j + 2\alpha\,.$$

Hence *the series of equivalence defined by* $\varkappa \equiv \alpha_j - 2\alpha$ *is an invariant series of* $\mathfrak{C}$. The linear series $|\varkappa|$, if effective, is called the *canonical series* of $\mathfrak{C}$.

14. The genus. Rational and elliptic curves. To calculate the order of $|\varkappa|$ we take $\mathfrak{C}$ to be a plane curve of order n, with d nodes, and consider the series g_n^1 cut on $\mathfrak{C}$ by a pencil of lines. The order of the corresponding Jacobian set is equal to the rank m of $\mathfrak{C}$ (§ 6), evidently given by the formula $m = n(n-1) - 2d$. Thus the required order is

$$n(n-1) - 2d - 2n = 2p - 2\,,\ \text{say.}$$

It follows that *the number p, defined by the relation*

$$p = \tfrac{1}{2}(n-1)(n-2) - d\,, \tag{8}$$

is a birational invariant of $\mathfrak{C}$. This is called the *genus* of the curve; and it is understood that, if any particular birational transform of $\mathfrak{C}$ possesses singularities which are not ordinary, these are to be resolved before the genus is computed.

Since, for a line, $p = 0$, it follows that *every rational curve has genus zero.* Conversely, *any curve for which $p = 0$ is rational.* We observe also that the canonical sets of a curve are virtual if, and only if, the curve is rational.

Ex. Apart from the above numerical test there are geometrical tests for rationality. Thus a curve is obviously rational if it contains a g_1^1.

For a non-singular plane cubic we have $p = 1$. Conversely, *any curve of genus unity is birationally equivalent to a non-singular plane cubic.* Now the equation of the latter can be reduced to the standard form $x_2^2 x_0 = 4\, x_1^3 - g_2 x_1 x_0^2 - g_3 x_0^3$, where g_2 and g_3 are constants; it therefore admits the parametric representation $x_1 : x_2 : x_0 = \wp(u) : \wp'(u) : 1$, where the Weierstrassian elliptic function $\wp(u)$ is formed with the invariants g_2, g_3. For this reason curves of genus unity are called *elliptic.* Geometrically, they are characterised by the fact that their canonical series is the null series.

15. The Riemann-Roch theorem. The canonical curve. A linear series is called *special* or *non-special* according as its sets are or are not contained in the canonical series of the carrying curve; to each special series there is attached an *index of speciality*, i. e. the number of linearly independent canonical sets which contain a generic set of the series. The Riemann-Roch theorem expresses the freedom of any series in terms of this concept: precisely, *if g_n^r is any complete linear series on a curve of genus p, then $r = n - p + i$, where i is the index of speciality of the series.*

For the canonical series itself ($n = 2p - 2$, $i = 1$) we thus have $r = p - 1$; hence the genus of the plane curve $\mathfrak{C}$ of § 14 may be defined as the number of linearly independent adjoints of order $n - 3$; these are termed *canonical adjoints.*

An arithmetical property of adjoint curves. Suppose that the curve $\mathfrak{C}$ has equation $f(x_0, x_1, x_2) = 0$, where the coefficients of the polynomial f are indeterminates, defining a field of rationality K; then *the series cut on $\mathfrak{C}$ by adjoints of any assigned order is determinable in K.* This result, which has notable applications, is a simple consequence of the fact that any rational symmetric function of the roots of an algebraic equation is rationally expressible in terms of the coefficients.

Provided that $p > 2$, we can in general construct a projective model of the series $|\varkappa|$ by identifying the linear system (7) with that of the canonical adjoints, and writing $\varrho x_i = \varphi_i$ ($i = 0, 1, \ldots, p-1$). We thus obtain a curve of order $2p - 2$, normal in S_{p-1}, which is called a *canonical curve* of genus p.

In order to effect this construction we must suppose that $|\varkappa|$ is *simple*, i. e. that the canonical sets which pass through the generic point of our curve do not in consequence pass through any other point of the curve. Actually there is one (and only one) exceptional case, the so-called *hyperelliptic* curve of genus p. Such a curve is characterised by the property of containing a series g_2^1, and its canonical sets each consist of $p - 1$ sets of this series.

16. Geometry on a surface. The case $k = 2$, to which we now turn, presents various features already familiar from the study of curves; but there are also some surprising developments which have no analogues in the preceding theory. To begin with, any surface can be birationally transformed into a surface of $S_R(R \geqq 5)$ which is free from singularities. If this is projected generically on to a space S_3 we obtain a model F which possesses the following scheme of multiple points.

On F there is a *double curve* $\mathfrak{C}$ (possibly reducible) at the generic point of which F has two distinct sheets; on $\mathfrak{C}$ there is a number of cuspidal points (or *pinch-points*) at which the sheets coincide. There is also a number of points which are *triple* both for $\mathfrak{C}$ and for F; at each such point the branches of $\mathfrak{C}$ and the sheets of F are all distinct. A surface of S_3, even if it is not a projection, which is endowed with such a double curve, is said to possess *ordinary singularities*. In the sequel we shall suppose that all the surfaces under discussion are either non-singular or endowed with ordinary singularities.

17. Linear curve systems. The first basic concept in the birational geometry of surfaces is in strict analogy with that of linear series on a curve. We say that the primals (7) cut on a surface F a *linear system* of curves. Such a system may possess *base points* (possibly multiple), and its generic curve may be reducible, with *fixed parts* (possibly multiple); but it is important to note that *the generic curve of the system cannot have a variable multiple point* (BERTINI's first theorem). A linear system of freedom r is said to have *index* unity in virtue of the fact that just one curve of the system passes through r generic points of F. For $r = 1, 2, 3$, we have a *linear (rational) pencil, net* and *web* respectively.

Alongside a rational pencil of curves we have frequently to consider an *irrational pencil*; this is a system, of freedom and index 1, whose members are in birational correspondence with the points of an irrational curve.

Ex. The generators of a cone which project a plane curve of genus $p > 0$ constitute an irrational pencil (of genus p) of lines. Any birational transform of the cone will change this system into another irrational pencil (also of genus p) of rational curves.

BERTINI's second theorem, which follows easily from his first, states that, *if the variable part of a linear system is reducible, it must be composed (compounded) of a certain number of curves of a pencil, rational or irrational.*

The theory of linear systems may be built up either from the concept of rational functions on a surface, or from the notion of adjoint surfaces. (A surface *adjoint* to a surface F with ordinary singularities is one which passes simply through the double curve of F.) Operations with linear systems are carried out precisely as in §§ 9—12; thus, in order to make subtraction always possible, we introduce *virtual curves* and *systems of* (linear) *equivalence*, subject to the conventions laid down in §§ 11—12, but now applied to curves instead of to point sets.

18. Characters of a linear system. The unique complete linear system $|\mathfrak{C}|$ defined by a given irreducible curve $\mathfrak{C}$ on F will certainly possess base points at the multiple points (if any) of $\mathfrak{C}$ (§ 17); in addition, however, it may have *unassigned* base points arising from those already assigned.

Ex. The plane cubics passing through eight assigned points all pass through a ninth point.

In the general theory this phenomenon has to be taken into account; one way of dealing with it is to regard the unassigned base points as *virtually non-existent* and to omit them from the calculations.

In this section we shall suppose, for the sake of simplicity, that the linear systems considered have no base points whatever. Then, given a linear system $|\mathfrak{C}|$, of freedom $r \geqq 2$, with irreducible generic $\mathfrak{C}$, we define for it two basic characters: the *genus* $\pi(\mathfrak{C})$ and the *characteristic set* $(\mathfrak{C}^2)$, i. e. the set of points common to two generic members of $|\mathfrak{C}|$. The number $[\mathfrak{C}^2]$ of points in $(\mathfrak{C}^2)$ is called the *grade* or *intersection number* of $|\mathfrak{C}|$.

The system $|\mathfrak{C}|$ cuts on $\mathfrak{C}$ a series, called the *characteristic series* of $\mathfrak{C}$. This series may be *simple* or *compound*: in the latter case the totality of sets $(\mathfrak{C}^2)$ belong to an *involution* on F, i. e. an aggregate of sets with the property that any set which contains a generic point of F consequently contains a number of additional points.

In the case where $|\mathfrak{C}|$ has freedom $r \geqq 3$ and a simple characteristic series, the system may be mapped on the prime sections of a surface F' of S_r, precisely as in § 15; and F' will be a birational transform of F. But if the characteristic series is compound, the generic point of F' will instead correspond to a set of points belonging to the involution on F, i. e. F, F' will be in unirational correspondence (§ 8).

Consider next the sum $|\mathfrak{C} + \mathfrak{D}|$ of two linear systems $|\mathfrak{C}|$ and $|\mathfrak{D}|$; in this case the genus $\pi(\mathfrak{C} + \mathfrak{D})$ is given by NOETHER's formula,

$$\pi(\mathfrak{C} + \mathfrak{D}) = \pi(\mathfrak{C}) + \pi(\mathfrak{D}) + [\mathfrak{C}\mathfrak{D}] - 1\,, \tag{9}$$

where $[\mathfrak{C}\mathfrak{D}]$ is the intersection number of the two systems. The grade is given by

$$[(\mathfrak{C} + \mathfrak{D})^2] = [\mathfrak{C}^2] + [\mathfrak{D}^2] + 2\,[\mathfrak{C}\mathfrak{D}]\,. \tag{10}$$

These formulae can be extended successively to give the characters of the sum of any number of linear systems, or of any positive multiple of $|\mathfrak{C}|$. They may also be used to define the *virtual grade* $[\mathfrak{C}^2]$ of any isolated curve $\mathfrak{C}$; this is effected by considering any sufficiently ample system $|\mathfrak{C} + \mathfrak{D}|$ which contains $\mathfrak{C}$.

Finally, in the field of virtual systems, they can be applied to evaluate the characters of the system $|\mathfrak{C} - \mathfrak{D}|$, and hence of the system defined by any algebraic sum of effective curves.

Ex. It follows from (9) and (10) that the *null curve* has virtual genus 1 and virtual grade 0.

19. Canonical and pluricanonical systems. Any *general* net $|\mathfrak{C}|$, i. e. one which does not contain ∞^1 reducible members, possesses a well-defined Jacobian curve $\mathfrak{C}_j$, locus of the double points of the net (for the imposition of such a point amounts to a single condition). And if $|\mathfrak{C}|$ is a general system of freedom $r > 2$, i.e. one not containing ∞^{r-1} reducible members, the Jacobian curves of all nets extracted from $|\mathfrak{C}|$ are contained in a linear system $|\mathfrak{C}_j|$, called the *Jacobian system* of $|\mathfrak{C}|$. If instead the net $|\mathfrak{C}|$ possesses a simple fixed part $\mathfrak{D}$, this counts three times in the composite Jacobian curve.

It follows that, for any general linear systems $|\mathfrak{C}|$, $|\mathfrak{D}|$, we have the equivalences

$$(\mathfrak{C} + \mathfrak{D})_j \equiv \mathfrak{C}_j + 3\,\mathfrak{D} \equiv \mathfrak{D}_j + 3\,\mathfrak{C}\,.$$

Hence the curve (effective or virtual) defined by $\mathfrak{K} \equiv \mathfrak{C}_j - 3\,\mathfrak{C}$ is an *invariant* of F, called a *canonical curve* of F. The system $|\mathfrak{K}|$, when effective, is termed the *canonical system*, and its freedom is denoted by $p_g - 1$, where p_g is the *geometric genus* of F. When there is no effective canonical curve we write $p_g = 0$.

The positive multiples of $|\mathfrak{K}|$ are called *pluricanonical curves*; the system $|i\mathfrak{K}|$, if effective, is termed the *i-canonical system* and its freedom is $p_i - 1$, where p_i is the ith plurigenus or i-genus. Examples show that, when $p_g = 0$, it is possible to have $p_i > 0$, for some values of i (cf. § 20).

In the case where F is a surface in S_3, of order n, with ordinary singularities, and $\mathfrak{C}$ is a plane section, it is clear that curves of the system $|\mathfrak{C}_j|$ are cut on F by first polars; hence the system $|\mathfrak{K}|$, if effective, is cut on F by adjoint surfaces of order $n - 4$; these are called *canonical adjoints*.

With any system $|\mathfrak{C}|$ there is associated a system $|\mathfrak{C}'|$, termed *adjoint* to $|\mathfrak{C}|$, defined by $|\mathfrak{C}'| = |\mathfrak{K} + \mathfrak{C}|$. This has so far been shown only when $|\mathfrak{C}|$ is general in the sense defined, but the concept extends to all cases. It follows from § 13 that *the system* $|\mathfrak{C}'|$ *cuts sets of the canonical series on any* $\mathfrak{C}$.

Now the curve $\mathfrak{K}$ has in every case a virtual genus whose value may be computed according to the rules already laid down; this is called the *virtual linear genus* of F, and denoted by $p^{(1)}$. As we shall now see, this is only a *relative* invariant for F, since its value may be altered by a birational transformation which introduces exceptional elements.

Consider a net $|\mathfrak{C}|$ which has a *fundamental curve* $\mathfrak{J}$, i.e. a curve which has no variable intersections with the generic $\mathfrak{C}$; it follows that the curves of $|\mathfrak{C}|$ which pass through a point of $\mathfrak{J}$ must contain $\mathfrak{J}$ entirely, so that it forms part of $\mathfrak{C}_j$. In the case where $\mathfrak{J}$ is a simple component of $\mathfrak{C}_j$ it will also be in general a simple component of $\mathfrak{K}$. Such a curve is called *exceptional*; it may be transformed into the neighbourhood of a simple point on some birational transform of F (cf. § 8).

If F possesses an effective canonical or pluricanonical system it can obviously contain only a finite number of exceptional curves. The system $|\mathfrak{K}|$ is called *pure* or *impure* according as F is or is not free from such curves. Clearly these have no effect upon the characters p_g and p_i, which are therefore absolute invariants of F. But they do affect the value of $p^{(1)}$. Actually it is known that, if F possesses an effective canonical or pluricanonical system, all its exceptional curves can be removed by birational transformation of F. The *absolute linear genus* of F is then defined as the linear genus of any model of F which is free from exceptional curves.

20. The arithmetic genus. While p_g and p_i are geometric invariants of a surface we have seen that the genus of a curve, as defined in § 14, is a numerical invariant. The analogous character of F, called the *arithmetic genus* p_a, arises from the computed dimension $p_a - 1$ of the system $|\mathfrak{K}|$. In the case where F has order n and ordinary singularities, with a double curve (possibly reducible) of order m, virtual genus π, and with t triple points, p_a is given by the formula

$$p_a = \binom{n-1}{3} - m(n-4) + 2t + \pi - 1 . \qquad (11)$$

The difference $q = p_g - p_a$, which is always non-negative, is called the *irregularity* of F; F is said to be *regular* or *irregular* according as $q = 0$, or $q > 0$.

Ex. Consider the ENRIQUES surface, which is a sextic surface whose double curve consists of the six edges of a tetrahedron in S_3; the equation of such a surface is easily written (cf. p. 97). Since there is no canonical adjoint of order $n - 4$, $p_g = 0$; and we find that $p_a = 0$, also. But there exists an effective bicanonical curve (actually of order zero) since the tetrahedron itself constitutes the double of a canonical adjoint, and the intersection of the tetrahedron with the surface consists entirely of the double curve.

21. The RIEMANN-ROCH theorem for surfaces. Let $|\mathfrak{C}|$ be any system, irreducible or otherwise, with virtual grade n and virtual genus π; then the freedom r of $|\mathfrak{C}|$ is given by the inequality

$$r \geqq n - \pi + p_a + 1 - i , \qquad (12)$$

where i is the index of speciality of $|\mathfrak{C}|$, i. e. the number of linearly independent canonical curves of F which contain $\mathfrak{C}$. A system $|\mathfrak{C}|$ is said to be *regular* if $i = 0$, and if the equality sign holds in (12); the difference $r - (n - \pi + p_a + 1 - i)$ is called the *superabundance* of $|\mathfrak{C}|$.

22. Series of equivalence. Besides the concept of linear system there is a second notion of a linear aggregate which is likewise invariant for birational transformation of a surface F, namely that of point sets which belong to a *series of equivalence*. This theory, which is outlined on pp. 21—22, gives rise to *invariant series of equivalence* analogous to the canonical system, whose orders are (relative) numerical invariants of F. The most important of these is the SEVERI series (p. 22), of order $I + 4$, where I is known as the ZEUTHEN-SEGRE invariant of F. This is also given by NOETHER's relation

$$I + p^{(1)} = 12 p_a + 9 . \qquad (13)$$

23. Some important classes of surface. Of all types of surface, the rational is the simplest. Calculating its invariants from the plane model we find that $p_g = p_a = p_i = 0$. CASTELNUOVO has proved that a surface is rational provided that $p_a = p_2 = 0$.

The projective classification of the rational surfaces proceeds from the genus π of the generic curve section. If $\pi = 0$, the surface is either ruled or is a VERONESE quartic surface or one of its projections. If $\pi = 1$, the surface is either ruled or rational; in the latter case it is known as a DEL PEZZO surface, and may have any order from 3 to 9 inclusive.

Next in order we have the *scrollar* surfaces, which are birationally equivalent to ruled surfaces. For such surfaces we find that $p_g = p_i = 0$, while $p_a = -p$, where p is the genus of a curve section. ENRIQUES has shown that the rational and scrollar surfaces are characterised by the single condition $p_{12} = 0$.

Of the irregular surfaces which are not scrollar, perhaps the most interesting are the JACOBI and PICARD surfaces. The simplest type of JACOBI surface is that which maps the product of two elliptic curves (§ 7); the general type is definable as the surface which maps the pairs of points of a curve of genus 2. The PICARD surface is then definable as the surface which maps an involution of point sets on the JACOBI surface, the involution being without coincidences, i. e. the members of each set are always distinct.

From a function-theoretic point of view, the PICARD surface belongs to the same family of varieties as the elliptic curve (§ 14), for it admits a parametric representation by means of Abelian functions of genus 2 (hyperelliptic functions).

24. Algebraic curve systems. Theory of the base. While the members of a linear curve system can be set in birational correspondence with the points of a birational variety, those of an *algebraic* system are in general mapped by the points of an irrational variety. For a regular surface the distinction is unimportant, in view of the result that, on such a surface, *any algebraic system is contained in a linear system.* But an irregular surface carries algebraic systems which are not so contained, e. g. irrational pencils. For this type of surface we require to formulate a concept of *algebraic equivalence* which generalises that of linear equivalence, and which takes into account the idea of virtual curves.

It is found that, for this purpose, the notion of an algebraic system which is irreducible as an aggregate of *curves* is inadequate: we have to replace this by a *continuous system,* i. e. a system which is *irreducible as a totality of linear systems.* We then say that two curves $\mathfrak{C}_1$, $\mathfrak{C}_2$ are algebraically equivalent if, and only if, there exists a curve $\mathfrak{D}$ such that $\mathfrak{C}_1 + \mathfrak{D}$ and $\mathfrak{C}_2 + \mathfrak{D}$ belong to the same continuous system.

On any surface F (assumed to be non-singular or endowed with ordinary singularities) it is possible to determine a finite number of curves $\mathfrak{C}_1, \mathfrak{C}_2, \ldots, \mathfrak{C}_\varrho$ such that a convenient multiple of any other curve of F is algebraically equivalent to $\lambda_1 \mathfrak{C}_1 + \lambda_2 \mathfrak{C}_2 + \cdots + \lambda_\varrho \mathfrak{C}_\varrho$, where $\lambda_1, \lambda_2, \ldots$ are integers. The curves $\mathfrak{C}_1, \mathfrak{C}_2, \ldots$ are said to form a *base,* and ϱ is termed the *base number* of F.

The surface F is said to be endowed with *divisor* if it contains at least two curves $\mathfrak{C}_1, \mathfrak{C}_2$ with the property that certain multiples $m\mathfrak{C}_1, m\mathfrak{C}_2$ are algebraically equivalent (to a curve $\mathfrak{C}$, say) while $\mathfrak{C}_1$ and $\mathfrak{C}_2$ are algebraically disequivalent; the curve $\mathfrak{C}$ is then said to possess the divisors $\mathfrak{C}_1, \mathfrak{C}_2$. It follows from the theory of the base that, for certain curve systems of F, the number of possible modes of division attains a maximum: this is termed the *divisor* σ of F.

Ex. For any rational or scrollar surface, $\sigma = 1$; for the ENRIQUES surface (§ 20), $\sigma = 2$. This is perhaps the simplest example of a surface endowed with divisor.

Bibliography.

Books and Reports.

ANDREOTTI, A.: (a) Abelian varieties and their applications. Berlin (in preparation).

BAKER, H. F.: (a) Principles of geometry, VI. Cambridge 1933; (b) A locus with 25920 linear self-transformations. Cambridge 1946.

BALDASSARRI, M.: (a) Algebraic varieties. Berlin (in preparation).

CASTELNUOVO, G.: (a) Memorie scelte. Bologna 1937.

CONFORTO, F.: (a) Funzioni abeliane e matrici di RIEMANN. Roma 1942.

ENRIQUES, F.: (a) Le superficie algebriche. Bologna 1949.

— and O. CHISINI: (a) Teoria geometrica delle equazioni, III. Bologna 1924.

HODGE, W. V. D., and D. PEDOE: (a) Methods of algebraic geometry, II. Cambridge 1952.

LEFSCHETZ, S.: (a) Selected topics in algebraic geometry, chap. **17**. Washington 1928; (b) L'Analysis situs et la géométrie algébrique. Paris 1950.

PAINLEVÉ, P.: (a) Leçons sur la théorie analytique des équations différentielles. Paris 1897.

ROTH, L.: (a) Algebraic threefolds. Rend. di Mat. (5) **10**, 297 (1951).

SAMUEL, P.: (a) Méthodes d'algèbre abstraite en géométrie algébrique. Berlin 1955.

SEGRE, B.: (a) Arithmetical questions on algebraic varieties. London 1951; (b) Geometry upon an algebraic variety. Proc. Inter. Math. Congress 1954, I.

SEVERI, F.: (a) Serie, sistemi d'equivalenza, I. Roma 1942; (b) Funzioni quasi-abeliane. Roma 1947; (c) Memorie scelte. Bologna 1950; (d) Trattato di geometria algebrica. Bologna 1926.

SNYDER, V.: (a) The problem of the cubic variety in S_4. Bull. Amer. Math. Soc. **35**, 607 (1929).

ZARISKI, O.: (a) Algebraic surfaces. Berlin 1935.

Papers.

ALBANESE, G.: [1] Rend. Acc. Lincei (5) **33**, 179 (1924); [2] Rend. Acc. Lincei (5) **33**, 210 (1924); [3] Annali di Mat. (4) **4**, 153 (1927); [4] Atti Congr. Internaz. Bologna 1928, IV, 157. Bologna 1931.

ANDREOTTI, A.: [1] Mem. Ac. Roy. Belgique **27** (1952); [2] Acta Ac. Pont. Sci. **14**, 107 (1952).

D'AMICO, F.: [1] Atti Acc. Gioenia in Catania (4) **18** (1906).

APRILE, G.: [1] Rassegna di Mat. e Fis. **1**, 135 (1921).

ARCHBOLD, J. W.: [1] Proc. Lond. Math. Soc. (2) **47**, 101 (1941); (2] Proc. Camb. Phil. Soc. **29**, 484 (1933).

BABBAGE, D. W.: [1] Proc. Camb. Phil. Soc. **32**, 12 (1936).

BALDASSARRI, M.: [1] Rend. Sem. Mat. Padova **19**, 1 (1950); [2] Rend. Sem. Mat. Padova **20**, 135 (1951); [3] Rend. Acc. Lincei (8) **12**, 390 (1952).

BARKER, C. C. H.: [1] Journ. Lond. Math. Soc. **26**, 125 (1951); [2] ibid. **30**, 343 (1955).

BARSOTTI, I.: [1] Rend. Palermo (2) **2**, 236 (1953).

BENEDICTY, M.: [1] Rend. Acc. XL (3) **27**, 1 (1948).

Castelnuovo, G., and F. Enriques: [1] Ann. Ec. Norm. Sup. (3) **23**, 339 (1906); [2] Annali di Mat. (3) **6**, 165 (1901); [3] Comptes rendus **121**, 242 (1895).

Comessatti, A.: [1] Rend. Acc. Lincei (5) **22**, 270, 316, 361 (1913); [2] Rend Palermo **46**, 1 (1922); [3] Boll. Un. Mat. Ital. (2) **2**, 97 (1940).

Conforto, F.: [1] Rend. Acc. Italia (7) **2**, 268 (1941).

Dantoni, G: [1] Atti Acc. Italia **14**, 39 (1943); [2] Annali di Mat. (4) **24**, 177 (1945).

Du Val, P.: [1] Journ. Lond. Math. Soc. **28**, 1 (1953).

Eger, M.: [1] Ann. Ec. Norm. Sup. (3) **60**, 143 (1943).

Enriques, F.: [1] Math. Annalen **49**, 1 (1897); [2] Mem. Acc. XL **10**, 201 (1896); [3] Math. Annalen **52**, 449 (1899); [4] Math. Annalen **46**, 179 (1895); [5] Annali di Mat. (3) **20**, 109 (1913); [6] Rend. Acc. Lincei (5) **21**, 81 (1912); [7] Rend. Acc. Lincei (6) **16**, 540 (1932); [8] Atti Acc. Torino **29**, 275 (1894); [9] Rend. Acc. Lincei (5) **15**, 665 (1906);

— and G. Fano: [1] Annali di Mat. (2) **26**, 59 (1897).

Fano, G.: [1] Mem. Acc. Torino (2) **46**, 187 (1896); [2] Atti Ist. Veneto (7) **7**, 276 (1896); [3] Rend. Acc. Lincei (5) **8**, 562 (1899); [4] Atti Acc. Torino **39**, 597 (1904); [5] Atti Acc. Torino **43**, 973 (1908); [6] Atti Acc. Torino **44**, 633 (1909); [7] Atti Acc. Torino **50**, 1067 (1915); [8] Annali di Mat. (3) **24**, 49 (1915); [9] Rend. Acc. Lincei (6) **11**, 329 (1930); [10] Rend. Acc. Lincei (6) **15**, 3 (1932); [11] Scritti mat. offerti a L. Berzolari (Pavia 1936), 329; [12] Mem. Acc. Italia **8**, 23 (1937); [13] Mem. Acc. XL (3) **24**, 41 (1938); [14] Comm. Math. Helvetici **14**, 193 (1941); [15] Comm. Math. Helvetici **14**, 203 (1941); [16] Comm. Math. Helvetici **15**, 71 (1942); [17] Comm. Pont. Ac. Sci. **11**, 635 (1947); [18] Rend. Acc. Lincei (8) **6**, 151 (1949).

Franchetta, A.: [1] Rend. Acc. Italia (7) **2**, 282 (1941); [2] Rend. Acc. Lincei (8) **2**, 276 (1947).

Gaeta, F.: [1] Rend. Acc. XL (4) **2**, 1 (1951); [2] Annali di Mat. (4) 33, 91 (1952).

Gauthier, L.: [1] Bull. Soc. Roy. Liége **13**, 191 (1944).

Godeaux, L.: [1] Bull. Ac. Roy. Belgique (5) **19**, 134 (1933); [2] Colloque de géométrie algébrique, p. 177. Liége 1949.

Hall, R.: [1] Rend. di Mat. (5) **11**, 167 (1952); [2] Journ. Lond. Math. Soc. **29**, 419 (1954); [3] Journ. Lond. Math. Soc. **30** (1955).

Hilbert, D.: [1] Math. Annalen **36**, 473 (1890).

Hirzebruch, F.: [1] Arithmetic genera and the theorem of Riemann-Roch for algebraic varieties. Princeton Univ. mimeographed notes 1953; [2] The index of an oriented manifold and the Todd genus for an almost complex manifold. Princeton Univ. mimeographed notes 1953; [3] Todd arithmetic genus for almost complex manifolds. Princeton Univ. mimeographed notes 1953; [4] Proc. Nat. Ac. Sci. **39**, 951 (1953); [5] Proc. Nat. Ac. Sci. **40**, 110 (1954).

Hodge, W. V. D.: [1] Journ. Lond. Math. Soc. **30**, 291 (1955).

Humbert, G.: [1] Journ. de Math. (4) **10**, 197 (1894).

Igusa, J.: [1] Journ. Math. Soc. Japan **3**, 215 (1951).

Jongmans, F.: [1] Mem. Soc. Roy. Liége **7**, 1 (1947); [2] Bull. Ac. Roy. Belgique (5) **29**, 766, 823 (1943); [3] Mem. Ac. Roy. Belgique (2) **23** (1949).

Kodaira, K.: [1] Amer. Journ. Math. **73**, 813 (1951); [2] Annals of Math. (2) **56**, 298 (1952); [3] Proc. Nat. Ac. Sci. **38**, 522 (1952); [4] Annals of Math. (2) **59**, 86 (1954). [5] Annals of Math. (2) **60**, 28 (1954).

— and D. C. Spencer: [1] Proc. Nat. Ac. Sci. **39**, 641 (1953).

Lefschetz, S.: [1] Trans. Amer. Math. Soc. **22**, 327 (1921); [2] Trans. Amer. Math. Soc. **22**, 407 (1921).

Maroni, A.: [1] Annali di Mat. (4) **5**, 185 (1928).

Matsusaka, T.: [1] Nat. Sci. Rep. Ochanomizu Univ. **4**, 164 (1954).

Maxwell, E. A.: [1] Proc. Camb. Phil. Soc. **32**, 185 (1936).

MONTESANO, D.: [1] Rend. Acc. Napoli (3) **1**, 93, 155 (1895).

MORIN, U.: [1] Rend. Acc. Lincei (6) **24**, 191 (1936); [2] Rend. Acc. Lincei (6) **27**, 330 (1938); [3] Rend. Sem. Mat. Padova **9**, 1 (1938); [4] Annali di Mat. (4) **18**, 147 (1939); [5] Annali di Mat. (4) **19**, 257 (1940); [6] Annali di Mat. (4) **21**, 113 (1942); [7] Atti Congr. Un. Mat. Ital., p. 298. Bologna 1940; [8] Rend. Sem. Mat. Padova **21**, 298 (1952).

MUHLY, H. T., and O. ZARISKI: [1] Trans. Amer. Math. Soc. **69**, 78 (1950).

NÉRON, A.: [1] Bull. Soc. Math. France **80**, 101 (1952).

NOETHER, M.: [1] Math. Annalen **3**, 161 (1871).

NOLLET, L.: [1] Bull. Ac. Roy. Belgique (5) **36**, 897 (1950); [2] ibid. (5) **40**, 914 (1954).

D'ORGEVAL, B.: [1] Bull. Ac. Roy. Belgique (5) **36**, 302 (1950).

PANNELLI, M.: [1] Rend. Acc. Lincei (5) **15**, 483 (1906); [2] Rend. Acc. Lincei (5) **15**, 619 (1906).

PREDONZAN, A.: [1] Rend. Acc. Lincei (8) **5**, 238 (1948); [2] Rend. Sem. Mat. Padova **18**, 163 (1949); [3] Rend. Sem. Mat. Padova **21**, 278, 335 (1952); [4] Rend. Sem. Mat. Padova **23**, 245 (1954); [5] Rend. Sem. Mat. Padova **23**, 127 (1954).

ROSENBLATT, A.: [1] Atti Congr. Internaz. Bologna 1928 IV, 123. Bologna 1931.

ROTH, L.: [1] Proc. Lond. Math. Soc. (2) **35**, 540 (1933); [2] Proc. Lond. Math. Soc. (2) **39**, 334 (1935); [3] Proc. Camb. Phil. Soc. **33**, 188 (1937); [4] Rend. Acc. Lincei (8) **10**, 468 (1951); [5] Rend. Acc. Lincei (6) **21**, 314 (1935); [6] Proc. Lond. Math. Soc. (2) **40**, 217 (1936); [7] Proc. Camb. Phil. Soc. **29**, 184 (1933); [8] Proc. Lond. Math. Soc. (2) **30**, 118 (1930); [9] Proc. Lond. Math. Soc. (2) **30**, 305 (1930); [10] Proc. Lond. Math. Soc. (2) **35**, 249 (1933); [11] Proc. Camb. Phil. Soc. **31**, 508 (1935); [12] Journ. Lond. Math. Soc. **11**, 147 (1936); [13] Proc. Camb. Phil. Soc. **33**, 301 (1937); [14] Proc. Lond. Math. Soc. (2) **40**, 365 (1936); [15] Annali di Mat. (4) **27**, 115 (1948); [16] Annali di Mat. (4) **29**, 91 (1949); [17] Proc. Camb. Phil. Soc. **46**, 419 (1950); [18] Rend. Acc. Lincei (8) **9**, 62 (1950); [19] Rend. Acc. Lincei (8) **9**, 246 (1950); [20] Rend. Acc. Lincei (8) **10**, 19 (1951); [21] Proc. Camb. Phil. Soc. **47**, 496 (1951); [22] Rend. di Mat. (5) **10**, 96 (1951); [23] Proc. Camb. Phil. Soc. **48**, 233 (1952); [24] Rend. Acc. Lincei (8) **12**, 66 (1952); [25] Rend. Acc. Lincei (8) **12**, 265 (1952); [26] Atti Congr. Un. Mat. Ital. 1951, II, 434. Roma 1953; [27] Rend. Acc. Lincei (8) **12**, 401 (1952); [28] Annali di Mat. (4) **34**, 247 (1953); [29] Rend. Palermo (2) **2**, 141 (1953); [30] Proc. Camb. Phil. Soc. **49**, 397 (1953); [31] Rend. di Mat. (5) **12**, 387 (1953); [32] Rend. Acc. Lincei (8) **15**, 376 (1953); [33] Proc. Camb. Phil. Soc. **50**, 360 (1954); [34] Rend. Sem. Mat. Padova **23**, 277 (1954); [35] Rend. di Mat. (5) **13**, 30 (1954); [36] Annali di Mat. (4) **38**, 281 (1955); [37] Rend. Sem. Mat. Milano **26** (1955); [38] Rend. Sem. Mat. Torino **14** (1955).

SCORZA, G.: [1] Annali di Mat. (3) **15**, 217 (1908).

SCOTT, D. B.: [1] Proc. Camb. Phil. Soc. **42**, 229 (1946).

SEGRE, B.: [1] Mem. Acc. Italia **5**, 479 (1934); [2] Mem. Ac. Roy. Belgique (2) **14**, 1 (1936); [3] Boll. Un. Mat. Ital. (1) **13**, 93 (1934); [4] Proc. Camb. Phil. Soc. **38**, 368 (1942); [5] Rend. Acc. Lincei (8) **3**, 411 (1947); [6] Colloque d'algèbre et de théorie des nombres, p. 135. Paris 1949; [7] Annali di Mat. (4) **35**, 1 (1953); [8] Annali di Mat. (4) **37**, 139 (1954); [9] Rend. di Mat. (5) **13**, 75 (1954); [10] Rend. Acc. Lincei (8) **10**, 94 (1951); [11] Annali di Mat. (4) **33**, 5 (1952).

SEMPLE, J. G.: [1] Proc. Camb. Phil. Soc. **25**, 145 (1929).

SEVERI, F.: [1] Rend. Palermo **28**, 33 (1909); [2] Mem. Acc. Torino (2) **52**, 61 (1903); [3] Rend. Palermo **17**, 73 (1903); [4] Rend. Acc. Lincei (5) **15**, 691 (1906); [5] Atti Ist. Veneto **65**, 625 (1906); [6] Rend. Acc. Lincei (5) **20**, 537 (1911); [7] Mem. Acc. Italia **5**, 206 (1934); [8] Rend. Acc. Italia **3**, 548 (1942);

[9] Annali di Mat. (4) **21**, 1 (1942); [10] Annali di Mat. (4) **25**, 1 (1946); [11] Annali di Mat. (4) **32**, 1 (1951); [12] Convegno internaz. di geometria differenziale, p. 21. Roma 1954; [13] Rend. Acc. Lincei (8) **18**, 131 (1955).

SNYDER, V.: [1] Rend. Palermo **38**, 344 (1914); [2] Giorn. di Mat. **61**, 125 (1923).

TAGG, E. D.: [1] Journ. Lond. Math. Soc. **14**, 216 (1939).

TODD, J. A.: [1] Proc. Lond. Math. Soc. (2) **43**, 190 (1937); [2] Journ. Lond. Math. Soc. **9**, 206 (1934); [3] Proc. Camb. Phil. Soc. **36**, 28 (1940); [4] Proc. Camb. Phil. Soc. **33**, 425 (1937); [5] Proc. Lond. Math. Soc. (2) **43**, 127 (1937); [6] Proc. Lond. Math. Soc. (2) **45**, 410 (1939); [7] Proc. Lond. Math. Soc. (2) **46**, 199 (1940); [8] Proc. Edinb. Math. Soc. (2) **5**, 117 (1938); [9] Proc. Camb. Phil. Soc. **34**, 144 (1938); [10] Proc. Lond. Math. Soc. (2) **47**, 81 (1941); [11] Quart. Journ. Math. **6**, 128 (1935); [12] Quart. Journ. Math. **7**, 168 (1936); [13] Journ. Lond. Math. Soc. **10**, 194 (1935); [14] Proc. Lond. Math. Soc. (2) **42**, 316 (1937); [15] Proc. Camb. Phil. Soc. **26**, 323 (1930); [16] Proc. Camb. Phil. Soc. **49**, 410 (1953).

— and E. A. MAXWELL: [1] Proc. Camb. Phil. Soc. 33, 438 (1937).

VESENTINI, E.: [1] Rend. Acc. Lincei (8) **16**, 199 (1954); [2] Rend. Acc. Lincei (8) **17**, 196 (1954).

ZARISKI, O.: [1] Annals of Math. (2) **45**, 472 (1944).

Index.

(A reference such as "A 1" is to section 1 of the Appendix)